Hartmut Bossel
System Zoo 1 Simulation Models
Elementary Systems, Physics, Engineering

About the author: Hartmut Bossel is Professor Emeritus of environmental systems analysis. He taught for many years at the University of California in Santa Barbara and the University of Kassel, Germany, where he was director of the Center for Environmental Systems Research until his retirement. He holds an engineering degree from the Technical University of Darmstadt, and a Ph.D. degree from the University of California at Berkeley. With a background in engineering, systems science, and mathematical modeling, he has led many research projects and future studies in different countries, developing computer simulation models and decision support systems in the areas of energy supply policy, global dynamics, orientation of behavior, agricultural policy, and forest dynamics and management. He has written numerous books on modeling and simulation of dynamic systems, social change and future paths, and has published widely in the scientific literature in several fields.

SYSTEM ZOO 1
SIMULATION MODELS

Elementary Systems, Physics, Engineering

Hartmut Bossel

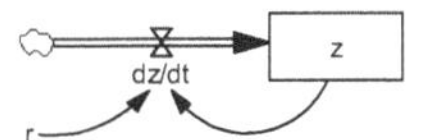

System Zoo 1 Simulation Models
© Hartmut Bossel 2007

Printed and published by
Books on Demand GmbH, Norderstedt, Germany

ISBN 978-3-8334-8422-3

Bibliografische Information Der Deutschen Bibliothek:
Die Deutsche Bibliothek verzeichnet diese Publikation
in der Deutschen Nationalbibliografie;
detaillierte bibliografische Daten sind im Internet über
http://dnb.ddb.de abrufbar.

Bibliographic information published by Die Deutsche Bibliothek:
Die Deutsche Bibliothek lists this publication
in the Deutsche Nationalbibliografie;
detailed bibliographic data are available in the Internet at
http://dnb.ddb.de

Information bibliographique de Die Deutsche Bibliothek:
Die Deutsche Bibliothek a répertorié cette publication
dans le Deutsche Nationalbibliografie;
les données bibliographiques détaillées peuvent être consultées
sur Internet à l'adresse http://dnb.ddb.de.

Preface

Daily life and the development of our world are determined by complex coupled dynamic systems: People, animals, plants, technology, enterprises, towns, clouds, forests. Although seemingly permanent in their outer appearance, they are constantly changing by mostly invisible processes, thereby also changing their environment. Knowledge about their possible dynamics is essential in many areas. The dynamic processes must be exposed and analyzed with the tools of systems analysis: with mathematical modeling and computer simulation.

The "System Zoo" is a collection of about one hundred simulation models[1] of complex dynamic systems from all areas of life in the departments: elementary systems, physics and engineering, climate and vegetation, ecosystems and resources, economy and society, global development. It is published in three volumes:

System Zoo 1 Simulation Models – Elementary Systems, Physics, Engineering
System Zoo 2 Simulation Models – Climate, Ecosystems, Resources
System Zoo 3 Simulation Models – Economy, Society, Development

All of these models are formulated according to globally accepted "system dynamics" standard, are documented in complete detail, have been thoroughly checked, are operational and can be operated with freely available simulation software with extensive processing possibilities. The model documentations are complemented by numerous exercises. The models are small enough to be implemented and utilized with modest effort, but most of them exhibit complex and often surprising behavior that is beyond intuitive assessment. Computer simulation provides a simple means of investigating and understanding the surprising variety of possible behaviors – a variety similar to that found in a zoo full of exotic animals. The "animals" in the System Zoo are grouped in the following six categories:

Part 1 "ELEMENTARY SYSTEMS" of the System Zoo documentation introduces smaller systems which are found as components in many complex systems, substantially determining their dynamics (like exponential and logistic growth, oscillations, delays etc.). This part is also an introduction to the practical side of modeling and simulation.

Part 2 "PHYSICS AND ENGINEERING" deals with an area in which the mathematical modeling of dynamic systems has arisen and in which simulations have always been quite important. The behavioral peculiarities of complex (nonlinear) systems such as limit cycles, attractors, multiple equilibrium points, and chaos are also examined here. More complex models are documented from the areas of control engineering, flight dynamics, and aerodynamics.

In Part 3 "CLIMATE AND VEGETATION" applications are introduced from the areas of climatology, the global CO_2 budget, photosynthesis and biomass production of plants, forest growth, and water, energy, and nutrient budgets of plant production in agriculture.

---

[1] Original publication: Bossel, H.: Systemzoo 1 – Elementarsysteme, Technik und Physik; Systemzoo 2 – Klima, Ökosysteme und Ressourcen; Systemzoo 3 – Wirtschaft, Gesellschaft und Entwicklung (3 vols.). Books on Demand, Norderstedt/Germany 2004. CD (German edition) with all models: Bossel, H.: Systemzoo. coTec Verlag, Rosenheim/Germany 2005.

Part 4 "Ecosystems and Resources" deals primarily with the dynamics resulting from interaction of plants, animals, and people with other organisms and environmental resources: by competition for food and nutrients, by use of renewable resources and exploitation of nonrenewable resources.

In Part 5 "Economy and Society" dynamic processes from this area are modeled and simulated: processes of production, supply chains, sales and consumption, competition for markets, personal life planning, unemployment, impacts of taxes on the development of commuter traffic and the economy, and finally also socio-psychological processes like escalation, dependence and aggression.

Part 6 "Global Development" presents simulation models which are relevant for the examination of longer-term societal developments with respect to population, housing, cost of living, pensions, state indebtedness, globalization, international competition, and global long-term development. Nonnumerical knowledge processing for the simulation of complex decision processes and impact assessment is also introduced.

The System Zoo is a collection of results and experience from a large number of research projects and many years of teaching, model development, and computer simulation. It is particularly well suited for courses in modeling and simulation, for class work as well as self-study, and for projects in schools, colleges, and universities. The three volumes of the System Zoo are complemented by a companion book providing the theoretical and practical foundations of mathematical modeling and computer simulation of systems: Hartmut Bossel: Systems and Models – Complexity, Dynamics, Evolution, Sustainability. Books on Demand, Norderstedt 2007 (ISBN 978-3-8334-8121-5). More detail on dynamic systems is provided in an earlier text: Hartmut Bossel: Modeling and Simulation. A K Peters, Wellesley MA 1994 (ISBN 1-56881-033-4).

Contents

Working with the System Zoo

Myriads of interconnected complex dynamic systems determine the development of our world and daily life. We recognize them in their particular and more or less permanent shape: people, animals, plants, forests, technology, enterprises, towns, states. But we hardly know and seldom recognize them as dynamic systems which are constantly changing by mostly invisible processes, thereby also changing their environment. This aspect usually escapes direct observation. It must be revealed using the tools of system analysis – in analogy to using the tools of x-ray analysis to reveal the structure and function of our bodies.

Animals and systems can be pictured and described in detail in encyclopedias and textbooks, but to get to know and understand their behavior we must observe them for some time under different conditions. Zoological gardens were created to provide people with the opportunity of observing animals – and in particular "strange" animals. In a zoo we can discover what books about animals cannot provide: the behavioral dynamics of a living being, often even in direct interaction with us. And a zoo offers a collection of very different animals with quite diverse behaviors in its different departments: mammals and birds, amphibians and fish, big and small animals, solitary and gregarious animals.

The three volumes of the System Zoo offer in their six chapters a collection of about one hundred simulation models of complex dynamic systems of all areas of life in the departments: elementary systems, physics and engineering, climate change and vegetation, ecosystems and resources, economy and society, global development. These simulation models are brought to life with easy-to-use simulation software. The models are described briefly at the beginning of each chapter. These descriptions should be read first to obtain an overview of the System Zoo and its residents.

The models and their simulation programs are all fully documented. They can be implemented on the PC with little effort using freely available interactive simulation software. It is the interactive work with these "system animals" that brings about often surprising insights about their dynamics and their often peculiar behavior. To save space, only one representative simulation run is usually documented for each model, but the spectrum of possible behaviors is always much richer than can be shown here. Each model description therefore mentions further interesting aspects which one should investigate to really understand the behavior of the system. Important: Most models are "generic" and are therefore valid also in completely different applications. Corresponding hints are provided in the respective model description.

As in visit to a zoo one should initially concentrate on systems of greatest personal interest. If you are new to the field of simulation, start your excursion with some of the simpler systems from the chapter ELEMENTARY SYSTEMS to familiarize yourself with the simulation software and its many processing possibilities. Detailed documentation and many teaching examples are also provided with the simulation software. The models in the different chapters are largely independent of each other. It therefore is not necessary (and not recommended) to complete the models one after the other. Rather, let your own interest and the joy of investigating and observing strange "animals" guide you.

The simulation models were developed using the software Vensim PLE (Personal Learning Environment) which is freely available in the internet for teaching purposes and private use (http://www.vensim.com). The symbolism ("system dynamics") used here is also used by other common simulation software like Stella (or ithink, http://www.hps-inc.com) and Powersim (http://www.powersim.com). The models presented here can therefore also be easily processed with these (and other) methods. All simulation models were tested in detail, in particular also with respect to the units of measurement (dimensions) used. In some models standardized nondimensional state variables normalized to "1" were used which, however, can also be easily converted to real dimensional formulations (cf. Bossel 2007 *Systems and Models*).

With few exceptions, the same notation is used for all model documentations: *Variables* are indicated by italics; small capitals are used for (mostly constant) PARAMETERS. State variables are shown as boxes in the system diagrams. Units of measurement are shown in [square brackets] in the program listings.

The software systems mentioned are particularly user-friendly. Their use can be learned quickly and easily. The software systems differ in small details, but operate in the same manner. To implement and compute the simulation models documented in the System Zoo the following steps have to be taken:

1. *Enter the simulation time parameters and save the model* under a name of its own. The time query usually appears in the first model frame; the details can be changed later.

2. *Place the system variables and parameters on the screen.* For this, select a corresponding button for (1) state variable, (2) rate of change, or (3) other system quantity, move the corresponding symbol to the desired place on the screen and place it there by mouse click. A rate (= valve symbol) is connected to a corresponding state variable by a "pipe". Enter the name of the variable or parameter.

3. *Connect the different quantities by influence arrows.* Select the button for the influence arrows, click on the "sender" quantity, draw the arrow to the "receiver" quantity, and put the arrow down by mouse click. If the quantities are too far apart in the simulation diagram, they can be connected using "shadow" or "ghost" variables (shown in the diagrams in <pointed> brackets).

4. *Quantify system quantities.* Click on the button for "equations", and then consecutively on each variable or parameter in the system diagram. A form appears with the name of the quantity (entered in Step 2) and the names of all input quantities connected to it (defined by the influence arrows as entered in Step 3). The mathematical function to be used for computing the quantity from the input quantities is entered in the form. For constant parameters the numerical values have to be entered.

5. *Start the simulation.* The program system checks completeness and correctness of the model formulation and reports possible (formal) errors. If these have been corrected, the simulation can be started using the "run" button. Euler-Cauchy integration is usually used but the more exact Runge-Kutta method (RK4) can also be chosen.

6. *Select the results and their presentation.* Every system quantity can be documented individually or together with other quantities using a variety of possible representations (diagrams, tables), in particular time diagrams or state space diagrams.

Simulation models of dynamic systems are mathematical models employing difference or differential equations which describe the (temporal) change of "state variables". It is not necessary to be familiar with this mathematical apparatus to work with simulation models and even to develop them. Working with the models of the System Zoo and learning from the methods employed there will provide valuable experience for developing your own models. Information on the theoretical and mathematical background of modeling and simulation of dynamic systems is found in an accompanying book (Bossel 2007 *Systems and Models*). This text presents elementary concepts like state equations, standardized and dimensionless quantities, equilibrium, oscillation, stability and instability, linearization, limit cycle, chaos etc. which are necessary for more intensive work and a deeper understanding of a system's dynamics.

Notes on working with the models: Although the model descriptions differ, every documentation follows the same sequence: Description of modeling task, simulation model, and major structural characteristics of the system; complete simulation diagram (sometimes several diagrams); complete listing of model equations; description of representative reference runs with time diagrams and state diagrams; hints at unusual features; exercises; references. Additional information on most models can be readily found in the internet.

All models are provided with default parameter settings which already demonstrate certain characteristic peculiarities. Suggestions for additional interesting investigations are provided with each model documentation. In addition to these model-specific suggestions the following general suggestions apply to all models:

1. *Start by examining the behavior of the reference run* (with the given default parameter settings) using the different modes of representation of results provided by the simulation software (e.g. time diagrams, state space representation, tables).

2. *Investigate the dependence of system development on "critical" parameters* mentioned in the documentation. It is recommended to make several runs with different values for a particular parameter, save the results, and compare these using diagrams or tables.

3. *Analyze the system in greater detail in parameter ranges where significant changes of system behavior can be observed* (e.g. stability/instability, equilibrium/collapse) or where other interesting effects appear.

4. *Investigate the global behavior* (for systems with two state variables) in the complete (relevant) state space for the reference case and/or interesting combinations of parameters (a supplementary module for the generation of state space diagrams is provided by model Z115 "State space diagram"). In particular, find the equilibrium points and determine stability or instability from the state trajectories.

5. *Calculate (analytically) the location of the equilibrium points* as function of the parameters using the state equations and the condition that the rates of change must disappear at the equilibrium points ($d\mathbf{z}/dt = \mathbf{0}$). Compare this with simulation results for the same parameter choice. (C.f. Bossel 2007 *Systems and Models* Ch. 2.8).

6. *Linearize the nonlinear state equations at the equilibrium points* and analyze the behavior of the corresponding linearized substitute system using the model of the linear oscillator (of same order; see models Z114, Z117) by substituting corresponding system parameters. Does the behavior of the original nonlinear system near the equilibrium points agree with the properties of the substitute linearized system and its eigenvalues? (see Bossel *Systems and Models* Ch. 2.8 for the theoretical background). (This suggestion refers primarily to two-dimensional systems and is intended for readers with some mathematical interest and background).

7. *Translate the (mostly) nondimensional generic models into simulation models for real systems* by correct dimensionalization of parameters and variables, and choice of suitable initial states and parameters. Compare the simulation results with experience and observations.

1
Elementary Systems

Overview

The behavior-defining structure of a dynamic system (i.e. the interactions of its components) can only rarely be recognized in the physical shape of the system. Systems may be entirely different physically and still have the same system structure resulting in the same behavior. System science is therefore a metascience (like mathematics) that applies to all areas of our reality and to all fields of science dealing with this reality. The System Zoo provides some examples from engineering, environment, economy and society which have nothing in common at first sight. It is only the system diagrams which frequently reveal common characteristics of system structure. Certain "elementary systems" are found as system modules in diverse combinations under very different conditions. Their characteristic behavior influences the behavior of the system in which they are embedded. Sometimes it has a decisive influence on the behavior of the complete system.

It is therefore important to be familiar with the fundamental behavior of such system modules, and to be able to assess their influence on the behavior of the complete system from the characteristics of their system structure and their coupling into the complete system. This part of the System Zoo deals with 17 relatively simple elementary systems which appear again and again in very different systems and contexts.

Z101 Simple integration. The most important system element, core of every dynamic system, is the combination of a state variable with its time rate of change. The prototype of this elementary system is the bathtub: The water quantity in the tub is the state variable (memory, stock). It changes by inflow and outflow. The change rate is positive if the faucet is open and the drain stopper closed. It is negative if the tap is closed and the stopper is drawn. This process can be described mathematically by integration over time: Starting with the initial condition (empty tub), inflow and outflow are integrated over time. The amount of water in the tub can thus be determined for every point in time. In this model different test functions are provided as rates of change, leading to corresponding time integrals for the system state.

Z102 System state and state change. Development of state with time completely depends on its change rate. In this model four frequently found change functions are used to examine their influence on the state: (1) Constant inflow, leading to linearly changing state; (2) inflow regulated as function of the state; (3) inflow changing as a predefined function of time; (4) inflow counteracting a "leak" in the system.

Z103 Exponential growth and decay. If the change of state is proportional to the current state and positive (positive feedback), the result is exponential growth. This is the well-known process of the accumulation of interest and compound interest: A savings account grows every year by a small fraction (the interest rate) of the amount already saved. Thus an initially small amount will become very large after a long time. If the rate of change is proportional to the current state and negative (negative feedback), the state decreases steadily. The less is still available, the lower the loss becomes. The state gradually approaches zero, but cannot become negative. Decay of radioactivity with a disintegration rate characteristic of the particular substance is such a process.

Z104 Exponential delay. The current level of a state variable is the result of past rates of change; it therefore acts as a memory of past inflows and outflows. Consider a state variable with a time-variant input. If by a negative feedback a fraction of the state is steadily lost as in the case of exponential decay, then this will primarily affect changes which occurred a long time ago while more recent changes are still more strongly reflected in the current state. A state variable with time-dependent input and such an "exponential leak" therefore "remembers" state changes with a time delay. This property is often used in system simulations to delay signals. However, the shape of signals changes somewhat in this process – they are smoothed. The exponential delay can therefore also be used for smoothing signals.

Z105 Time-dependent growth. If a state variable is coupled back to its change rate, the magnitude and sign of feedback determine whether and how fast the state grows or decays. The change rate is often a function of the system variables and can change strongly in the course of time. It may also be provided as a function of time affecting state development (e.g. as seasonal influence on plant growth). The model illustrates how the time function of a change rate can be entered as a table function or logic expression.

Z106 Simple population dynamics. Populations or stocks not only of organisms (plants, animals, people) but also of capital goods (houses, factories, vehicles) are subject to the processes of birth and death, disintegration and renewal. If births are more numerous than deaths, a population grows. However, births and deaths are proportional to the population level: The more people there are the more births and deaths will occur. The birth rate (or death rate) states by which percentage the population increases by births (or decreases by deaths) each year. Birth rate and death rate may change in the course of time (by birth control and medical progress or other factors). Such changes must be entered as scenarios in simulations of possible future developments.

Z107 Infection dynamics. The spreading of a contagious disease, a rumor, a new product or new knowledge, and many other processes of fundamental importance can be well described by a simple dynamic model with a single state variable. The population of those infected rises all the faster the more are already infected (and can pass on the infection) and the greater the population is of the still uninfected (the rest of the population). Once the greater part of the population has been infected, the infection rate decreases and the wave of infection will eventually die down. The spreading speed depends on contact frequency and infection probability for every contact.

Z108 Overloading a buffer. In the growth processes considered so far no upper limitation of the state variable was considered. In reality there are always limitations. A frequent limitation is overflow resulting from an overload: the overflowing bathtub or dam; flood peaks if soils cannot soak up heavy rain fall; short circuit at high voltage; and inappropriate behavioral response at high stress level. Typical of this process is a strong increase in the drain rate of the stored quantity (state variable) once a critical limit is exceeded. The process allows coping with excessive inflow rates (changes of state).

Z109 *Logistic growth with constant harvest.* Frequently a growth limitation arises (particularly in organisms and social processes) from feedback of population level on growth rate. For example, if initially only a few organisms populate a favorable terrain (or a new attractive product is introduced), the initial increase follows the maximum growth rate. Because more and more organisms now share the same food base (or fewer buyers are found who do not yet have the product), the growth rate gradually drops to zero. In this case the growth rate is controlled by the degree of saturation (with organisms or products) with respect to the carrying capacity (logistic growth). For example, the model describes approximately the growth of a fish population, and it also provides information about what may happen if fish are harvested at a given rate. Up to a certain critical harvest per year, the stock is not endangered: fishing would be sustainable. However, if harvesting increases beyond this critical limit, the fish population collapses inevitably and rapidly. Formulated as difference equation, the system exhibits chaotic behavior for high growth rates.

Z110 *Logistic growth with stock-dependent harvest.* A small change in a system can often cause its behavior to change radically. For example, if for a population subject to logistic growth care is taken that the harvest rate is proportional to the current stock (smaller harvest for smaller stock), the population will not collapse. This mechanism is typical of predator-prey relationships, where predation reduces the prey population and – because of lack of prey – also the predator population and hence predation.

Z111 *Density-dependent growth (Michaelis-Menten).* Certain biological and chemical saturation processes are correctly represented with a particular formulation of the saturation term. A "half saturation constant" determines the saturation behavior in this formulation. The process produces S-shaped growth similar to logistic growth.

Z112 *Double integration and exponential delay.* The double time integration of a state variable is particularly common in physical processes: The time rate of change of position of a mass equals its velocity, while the time rate of change of velocity equals its acceleration. If acceleration is integrated over time, one obtains the velocity as a function of time (for given initial velocity). If velocity is integrated over time, the position is obtained as a function of time (for a given initial position). If there exists negative feedback of the system states to their change rates, these "leaks" will produce "losses" (damping) of the dynamics and (a second order) exponential delay.

Z113 *Transition from one state to another.* Organisms or objects often remain for a while in a certain state before they change to a different one (and later perhaps into still another). *Examples*: Children will grow up, adults get old; butterfly eggs turn into caterpillars, caterpillars into pupae, pupae into butterflies; empty bottles are filled, packed, shipped, stored by the dealer, sold, put into the refrigerator of the buyer, and emptied. In such transitions, individuals are preserved (except for some deaths or bottles broken); they merely move from one state into a different one if certain criteria apply. The losses of one state (change rate: individuals per time unit moving to the next state) appear as gains (change rate: individuals per time unit entering the state) of the next state. In contrast to Z112 (double integration) the state variables retain the same dimension (individuals) after the integration over time. The transition between two states is the fundamental process of every population model.

Z114 Linear oscillator of second order. Systems capable of oscillation have enormous importance not only in all areas of engineering and physics but also in other areas. Oscillations can always appear if state variables are coupled by feedback loops, signals are delayed, and oscillations can therefore develop and amplify. In continuous systems (defined at any particular point in time, as are all natural systems) at least two state variables coupled by feedback are required for oscillation. Two state variables can be connected with altogether four different feedback loops. Depending on strength and sign of the feedbacks completely different, stable or unstable, periodic or aperiodic behavior is possible. These behavioral modes demonstrate the full spectrum of behaviors of which linear systems (also of higher order) are capable. Examples of oscillators are mechanical and electrical oscillation systems (e.g. vehicle suspension, electrical oscillating circuit), and periodic fluctuations in supply chains.

Z115 State space diagram. Nonlinear systems may show completely different behaviors in different domains of their state space. In contrast to linear systems they may have several (stable and unstable) equilibrium points, several domains of attraction, and exhibit limit cycles or chaotic behavior. To obtain a global overview of the system behavior for a given parameter constellation, the relevant state space must be investigated in its totality. This can be done for two-dimensional systems by plotting state trajectories in two-dimensional state space (phase portraits). For this task a supplemental program is introduced, which is used for analyzing the global behavior of several System Zoo models.

Z116 Triple integration and exponential delay. The exponential delay of third order is used in many simulations to compute the delay of signals (as is the first order delay Z104). In analogy to models Z104 and Z112, three integrators are arranged in sequence. In simulations with delays one has to remember that these are represented by state variables whose initial values must be provided (as in the case of other state variables). (They are often simply set to zero initially, but this is not always permissible!)

Z117 Linear oscillator of third order. Oscillators of order three and higher (having three and more linearly coupled state variables) play a role in many (mostly engineering) applications. However, their fundamental modes of behavior are identical with those of the linear oscillator of second order (Z114). The same mathematical approach applies, and the behavioral modes (stable, unstable, periodic, aperiodic) correspond.

Z101 Single integration

Simulation task

The core of every simulation of a dynamic system is the calculation of system states from the differential equations for the state changes per time unit, the "rates". This calculation corresponds to the integration of the rates of change over time, starting with given initial values for the states. This integration is carried out numerically in computer simulations (i.e. not analytically). Two widely used integration procedures which are found in every simulation software are the (simple but not very precise) Euler-Cauchy procedure and the (more complex but much more exact) Runge-Kutta procedure of 4th order (RK4) (Bossel 1994: 105-111, 391f).

A graphic example for integration over time is the computation of the contents of a reservoir (e.g. bathtub) as function of time and of the time-dependent inflows and outflows.

It goes without saying that the numerical integration must produce the same results as the well-known formulae of analytical integration, as far as these are applicable, for example

$$\int_0^T c\,dt = ct\Big|_0^T = cT$$

$$\int_0^T ct\,dt = \frac{1}{2}ct^2\Big|_0^T = \frac{1}{2}cT^2$$

$$\int_0^T \sin at\,dt = -\frac{1}{a}\cos at\Big|_0^T = \frac{1}{a}(-\cos aT + \cos 0) = \frac{1}{a}(1 - \cos aT)$$

In the following we model the elementary integration process and check agreement of the results of the numerical integration with the analytical formulae by integrating different test functions over time using the numerical model. Integrations of this type play an important role in every kind of state or inventory change (stocks, memory contents, accounts, populations etc.).

Simulation model

The model is shown in Figure Z101a. The model equations are presented in the following (dimensions are given in […]). In this case the rate of change is determined entirely by the input function. Depending on the parameters used this may represent a pulse train, a step function, a ramp function or a sine function (or a sum of these functions). Starting from the initial state, the corresponding (time-dependent) rate of change is integrated over time to produce the state as function of time.

Parameters and initial values
PULSE SEQUENCE = 1 [state/Day]
STEP FUNCTION = 0 [state/Day]
RAMP FUNCTION = 0 [state/Day]

SINE FUNCTION = 0 [state/Day]
FREQUENCY = 0.1 [1/Day]
INITIAL STATE = 0 [state]

Rates of change and state
Input function = PULSE SEQUENCE *50 *PULSE TRAIN(1, 0.02, 1, 20) +STEP
        FUNCTION*STEP(1, 1) +RAMP FUNCTION *RAMP(1, 1, 5) +SINE FUNCTION
        *SIN(2 *3.14159 *FREQUENCY *Time) [state/Day]
rate of change = Input function [state/Day]
state = INTEG (+rate of change, INITIAL STATE) [state]

Simulation time parameters
INITIAL TIME = 0 [Day]
FINAL TIME = 10 [Day]
TIME STEP = 0.02 [Day]
SAVEPER = TIME STEP [Day]

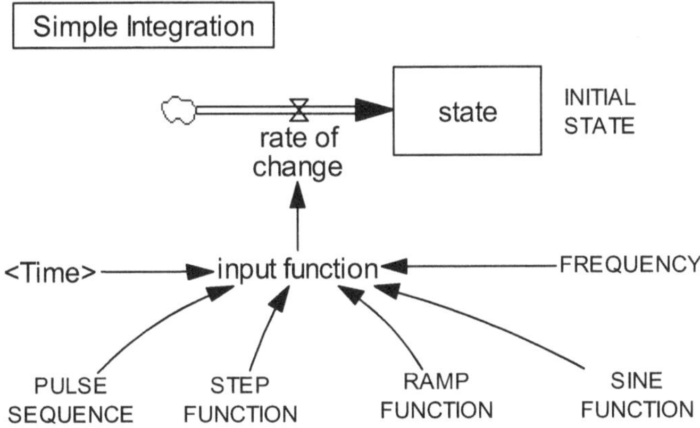

Figure Z101a: Simulation diagram for single integration.

Simulation results

Figure Z101b shows the results for the different input functions (for more on the test functions, see e.g. Bossel 1994, 162-168).

In the first case (Figure Z101b) the input function is a pulse train with pulse height 50 which starts at time $t = 1$ [day], each pulse having a duration of 0.02 [day] and being repeated at intervals of 1 [day] until time $t = 20$ [day] (values are given in the order of appearance in the equation for the input function). The pulse strength is = pulse height * pulse duration = 50 * 0.02 = 1. At each interval (of 1 day) a pulse increases the state suddenly by the amount = 1. A staircase function with a step height of 1 therefore describes the temporal function of the state.

In the second case (Figure Z101c) the input function is a (single) step function (a step function being the time integral of a pulse), which (with a step height of 1 at time

$t = 1$) produces a linear increase of the state with time. For $t = 10$ the state reaches a value of 9.

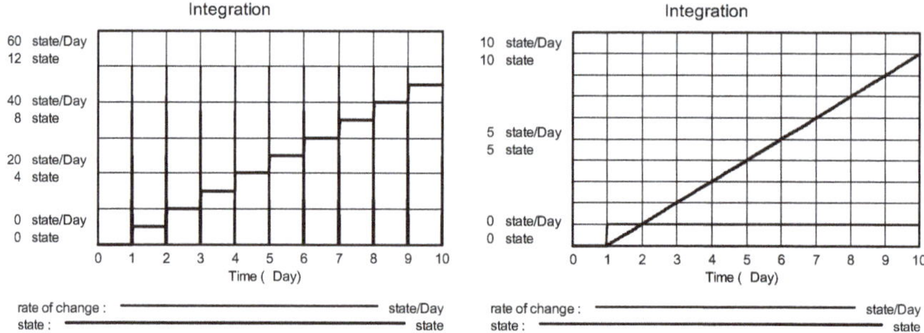

Figure Z101b: Integration of a pulse train yields a staircase function.
Figure Z101c: Integration of a step yields a ramp function.

In the third case (Figure Z101d) the input function is a (linear) ramp function (the time integral of the step function) with a slope of 1, which starts at time $t = 1$ and remains constant at the value (= 4) reached at time $t = 5$. This leads to an initial increase of the state as a quadratic function of time. After time $t = 5$ the change of state remains constant (= 4) and the state grows only linearly with time.

In the fourth case (Figure Z101e) the input function is a sine function sin at, where in this case (with the given default values) $a = 2 \pi (0.1) = 0.2 \pi$. With the initial state $z = 0$ for $t = 0$ the integration formula above produces for $t = 5$ a state value $z(5) = 1/(0.2 \pi) (1 - \cos (5 * 0.2 \pi) = 10/\pi = 3.183$. The simulation produces the same value for $t = 5$.

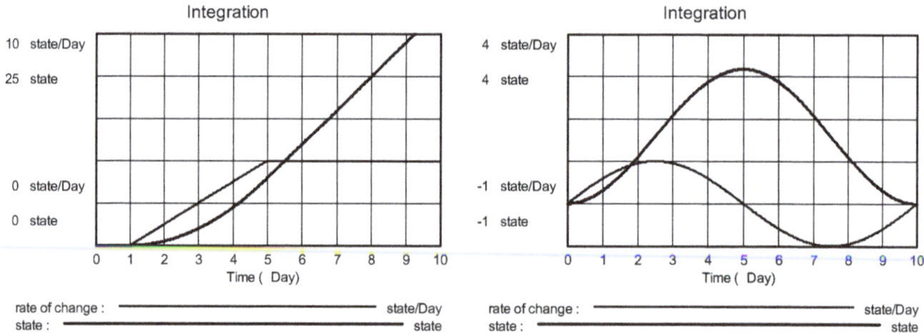

Figure Z101d: Integration of a ramp yields a quadratic function.
Figure Z101e: Integration of a sine function produces a cosine function.

We note the following from this simulation:
1. The integration "smoothes" the input function. This becomes particularly obvious for the pulse and the step function. Fluctuations in inflows or outflows are therefore partially damped and smoothed by reservoirs, stocks and other storage processes (i.e. integrators).

2. A positive sign of the input function leads to increases, a negative sign to losses in the stock (the state quantity).
3. The input function can be an arbitrary function of time.

Exercises

1. Confirm the correct numerical integration also for other input functions by comparing the simulation results with the results from analytical integration.
2. Examine the influence of the chosen computation step width (TIME STEP) on the precision of the result when integrating the sine function. What is the minimum number of computation steps that should be used in the integration over one period of the sine oscillation to keep the computation error below 1% if (a) Euler-Cauchy integration, and (b) Runge-Kutta integration is used?

Z102 System state and state change

Simulation task

The computation of a (time variable) state from its initial state and its rate of change can be easily understood by considering a simple physical process: the changing amount of material in a container as a consequence of (often time variable) inflows and outflows. A simple example is the filling process of a bathtub. It can be used to demonstrate some dynamic effects which also play an important role again and again in complex systems of completely different nature. We deal here with four simple systems, for which the inflows and outflows of a reservoir must be modeled in different ways.

Z102A Constant inflow, linear increase

The simulation diagram Z102Aa and the following program statements represent the constant inflow into a container. Initially the container is empty. At time $t = 0$ a constant inflow of 10 [liter/minute] starts. As a consequence, the level rises linearly and reaches a value of 100 [liter] after 10 minutes, as shown in the simulation results in Figure Z102Ab.

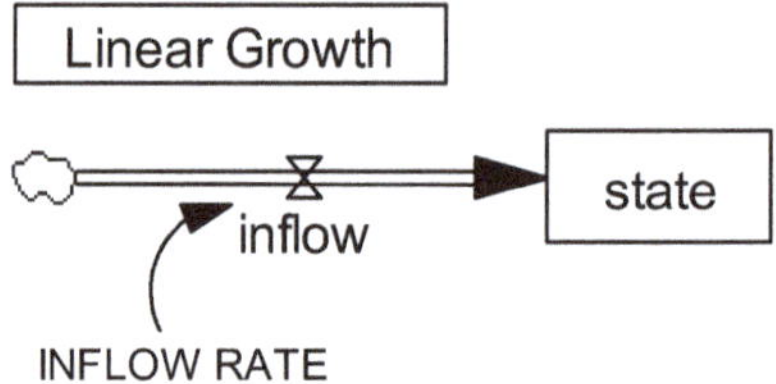

Figure Z102Aa: Simulation diagram for simple inflow.

Parameter
INFLOW RATE = 10 [Liter/Minute]

Dynamics
inflow = INFLOW RATE [Liter/Minute]
state = INTEG (inflow, 0) [Liter]

Simulation time parameters
INITIAL TIME = 0 [Minute]
FINAL TIME = 10 [Minute]
TIME STEP = 0.01 [Minute]

Z102B Inflow dependent on stock level

In the previous example it is implicitly assumed that the container does not overflow because it is either large or the overflow level is not reached during the simulation period. However, real containers have a finite volume, and this fact must be taken into account in models to avoid unrealistic conclusions.

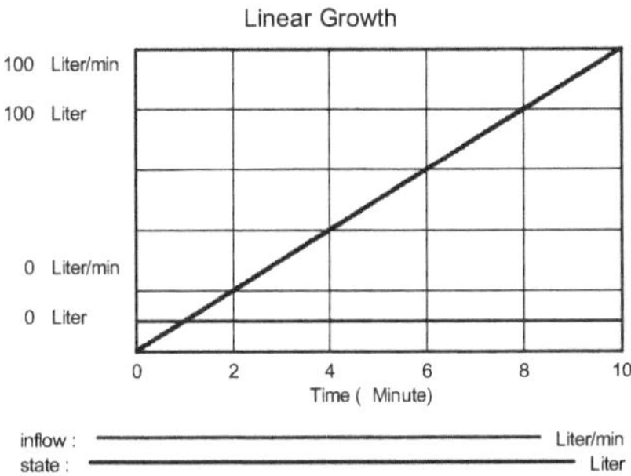

Figure Z102Ab: At constant inflow the level increases linearly

In the simulation diagram Z102Ba and in the following program statements the filling dynamics of a container (bathtub) having a limited BATHTUB VOLUME (of 100 [liter]) is simulated. In addition to its restricted capacity the container also has a leaking drain causing a constant loss OUTFLOW RATE. Since the *inflow* is larger than the *outflow*, the tub will gradually fill until the *state* is equal to BATHTUB VOLUME. After that the *inflow rate* is set to the value of the OUTFLOW RATE so that the tub remains "full" despite the constant leakage. Figure Z102Bb shows the corresponding simulation results.

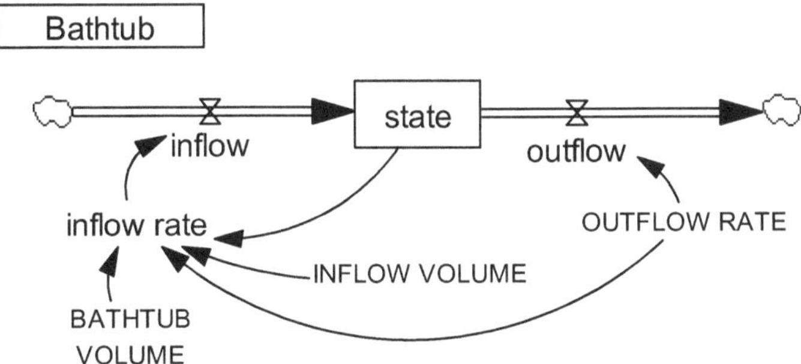

Figure Z102Ba: Simulation diagram for inflow depending on the level.

Parameters
BATHTUB VOLUME = 100 [Liter]
INFLOW VOLUME = 20 [Liter/Minute]
OUTFLOW RATE = 5 [Liter/Minute]

Dynamics
inflow rate = IF THEN ELSE (state < BATHTUB VOLUME, INFLOW VOLUME, OUT-
        FLOW RATE) [Liter/Minute]
inflow = inflow rate [Liter/Minute]
outflow = OUTFLOW RATE [Liter/Minute]
state = INTEG (inflow -outflow, 0) [Liter]

Simulation time parameters
INITIAL TIME = 0 [Minute]
FINAL TIME = 10 [Minute]
TIME STEP = 0.01 [Minute]

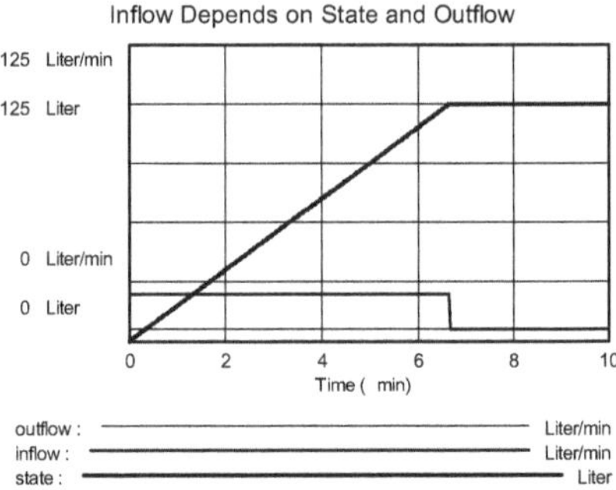

Figure Z102Bb: When the tub is full the net inflow becomes zero.

Z102C Time-dependent drain

In many applications the rates of change of state variables are determined by (empirical) functions of time, such as seasonal insolation, temperature, and precipitation. Such temporal functions must be entered into the simulation model as table functions (e.g. Bossel 1994: 93-94, 163-164). In most simulation software table functions can be easily entered and changed by drawing them directly on the computer screen.

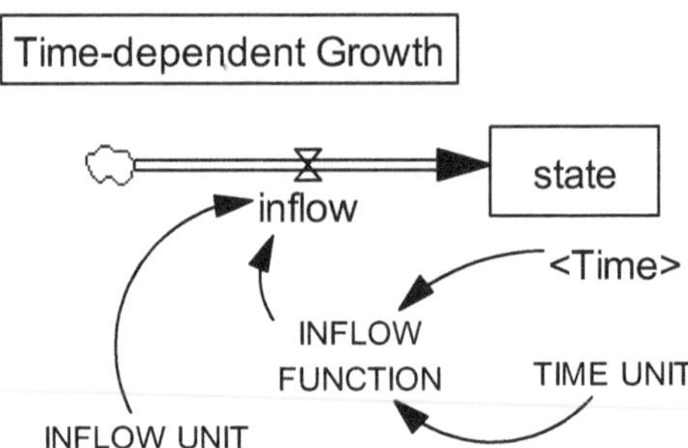

Figure Z102Ca: Simulation diagram for time-dependent input defined by a table function.

In simulation diagram Z102Ca and the following program statements the *inflow* is controlled by the time-dependent INFLOW FUNCTION, which was defined when writing the model, but which can also be changed for further simulation runs to examine its influence on the model behavior. Figure Z102Cb shows the simulation result. Here again it becomes obvious that an irregular input rate is smoothed to a much more regular time function for the state by the integration process. The temporal function used here is merely an example – the user should experiment with different inputs.

Parameters
TIME UNIT = 1 [Minute]
INFLOW UNIT = 1 [Liter/Minute]
INFLOW FUNCTION = WITH LOOKUP (Time /TIME UNIT, ([(0, 0) -(10, 10)], (0, 0), (0.6, 1.8), (1.3, 3.8), (1.8, 7.2), (2.2, 8.9), (2.9, 9.3), (3.3, 6.4), (3.6, 3.8), (4.3, 1.9), (4.8, 2.2), (5.6, 4.3), (6.1, 6.5), (6.3, 9), (7.1, 9.1), (7.4, 7.3), (7.7, 4.4), (8.2, 1.5), (9.5, 0), (10, 0))) [1]
Note: TIME UNIT is introduced to define a dimensionless independent variable in the table function. This is not really necessary, but it prevents a 'warning' resulting from the dimension check using the VenPLE-Software.

Dynamics
inflow = INFLOW FUNCTION *INFLOW UNIT [Liter/Minute]
state = INTEG (inflow, 0) [Liter]

Simulation time parameters
FINAL TIME = 10 [Minute]
INITIAL TIME = 0 [Minute]
TIME STEP = 0.01 [Minute]

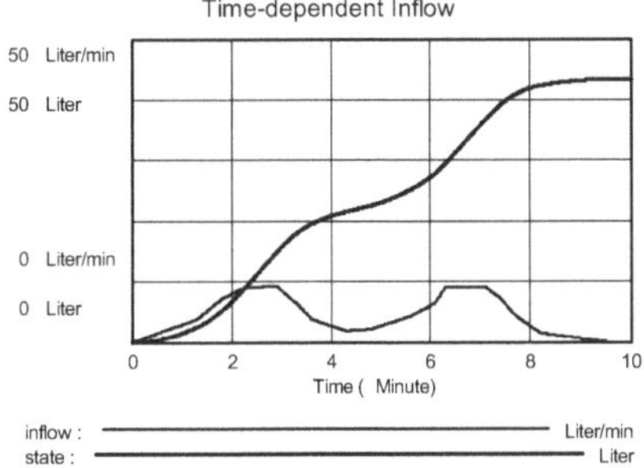

Figure Z102Cb: The state increases according to the time-variable inflow.

Z102D Container with leak

In many practical cases state variables have "leakages" which increase in proportion to increasing state levels. As an example, the outflow leaking from a container will increase as the water level rises and water pressure mounts. The process plays a role also in physically completely different contexts, e.g. in the natural absorption of pollutants in the soil.

In the simulation diagram of Figure Z102Da the stock dynamics is simulated for a container which receives a constant *inflow* of 20 [liter/minute] starting at time *t* = 1 [minute]. At the same time the container loses an amount of *outflow* which is proportional to its current *state* level. The simulation shows that the *state* increases continuously until the constant *inflow* can just compensate for the state-dependent *outflow* (Figure Z102Db). With the given parameters the *state* in the container cannot increase beyond 40 [liters]. At this point a flow equilibrium between *inflow* and *outflow* is established; subsequently the *state* remains constant.

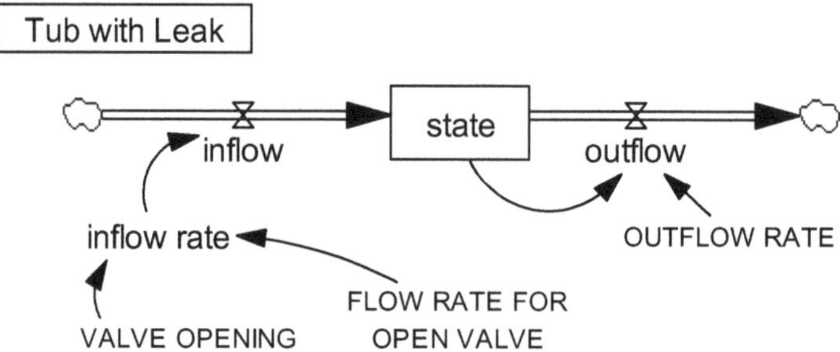

Figure Z102Da: Simulation diagram for a container with leak.

Parameters
FLOW RATE FOR OPEN VALVE = 20 [Liter/Minute]
VALVE OPENING = STEP (1, 1) [1]
OUTFLOW RATE = 0.5 [1/Minute]

Dynamics
inflow rate = FLOW RATE FOR OPEN VALVE *VALVE OPENING [Liter/Minute]
inflow = inflow rate [Liter/Minute]
outflow = OUTFLOW RATE *state [Liter/Minute]
state = INTEG (inflow -outflow, 0) [Liter]

Simulation time parameters
FINAL TIME = 10 [Minute]
INITIAL TIME = 0 [Minute]
TIME STEP = 0.01 [Minute]

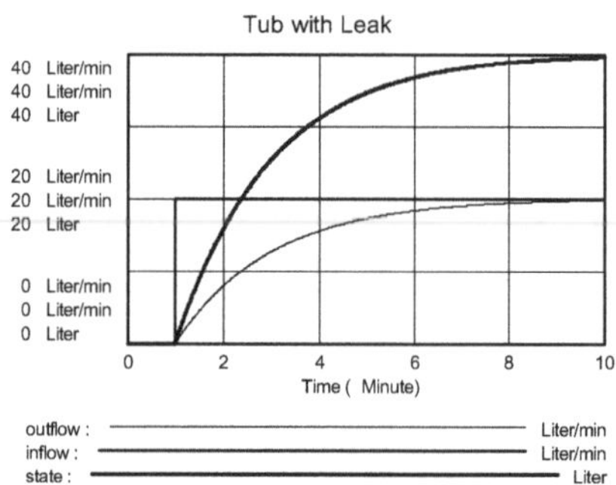

Figure Z102Db: The state remains unchanged if inflow = outflow.

Exercises

1. Change the different model parameters, and observe their effect on the simulation results. Make sure that you understand the respective process and the consequences of the parameter changes.

2. Enter table functions for the (positive and negative) inflow that produce (at the end of the simulation) again exactly the initial state value. Which condition must obviously apply to the relationship of the positive and negative shares of the time integrals of the inflow?

3. Formulate the filling process for the container with leak as a mathematical expression and show that the time dependence for the state variable is given by an exponential function (exp at = e^{at}). Determine analytically the equilibrium value for the state level.

4. Develop a simulation model for the calculation of a cattle stock over several years as a function of the number of calves (proportional to cattle stock) and number of cattle slaughtered (provided as a table function of time).

Z103 Exponential growth and decay

Simulation task

In many processes the change of state depends on the state itself: A large rabbit population will produce many offspring; the number of individuals will increase faster than for a small population. The volume of air escaping per time from a balloon will decrease as the balloon deflates. A large national economy can create more new infrastructure than a small one. As the amount of a radioactive substance decreases by radioactive decay, radiation per time also decreases.

In these systems the current state determines the rate of change of the state, i.e. the state is coupled back to itself (feedback). Without any external influence a self-generated dynamic process takes place. If by such a process a (positive) state value causes an increase of the state, this is referred to as positive feedback. Correspondingly, negative feedback would cause a reduction of the state level. Positive feedback is therefore associated with growth processes, while negative feedback is associated with reduction or decay of state variables. Negative feedback is often beneficial as it prevents uncontrolled growth.

Simulation model

The simulation model for this process is shown in Figure Z103a and the following model statements. The *change of state* is proportional to the *state*. The proportionality factor is the (specific) RATE OF CHANGE. It can be large or small, positive or negative. The sign determines whether it is a growth or a decay process. The (specific) RATE OF CHANGE always has the dimension [1/time unit]. Correspondingly the change of state always has the dimension [state unit / time unit].

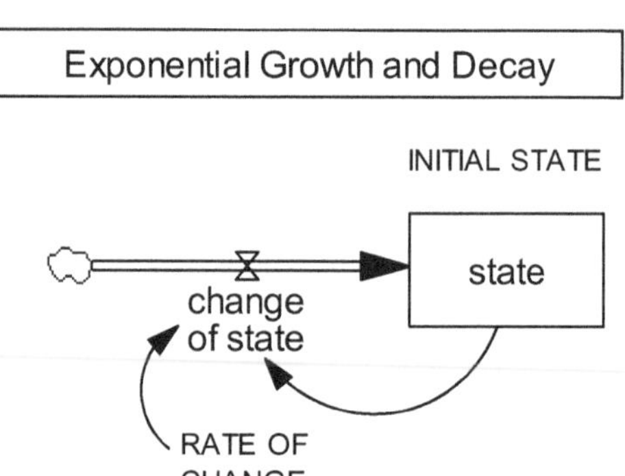

Figure Z103a: Simulation diagram for exponential growth or decay.

Parameters and initial state
RATE OF CHANGE = -1 [1/Year]
INITIAL STATE = 1 [amount]

Dynamics
state = INTEG (+change of state, INITIAL STATE) [amount]
change of state = RATE OF CHANGE *state [amount/Year]

Simulation time parameters
FINAL TIME = 10 [Year]
INITIAL TIME = 0 [Year]
TIME STEP = 0.02 [Year]

Simulation results

Figure Z103b shows the result for the particular parameter choice. In the case of strong negative feedback the state disappears almost completely in short time. The decay follows an exponential function of the time with the change rate $r = -1$:

$$x(t) = x_0\, e^{rt} = 1 \cdot e^{-1 \cdot t}$$

For $t = 1$ one obtains $x\,(1) = 1/e = 0.3678$, which is confirmed by the simulation result. If the formula is differentiated with respect to time, the differential equation for the change of state follows, i.e. *change of state* = RATE OF CHANGE · *state*.

$$\frac{dx}{dt} = r\,(x_0 e^{rt}) = rx$$

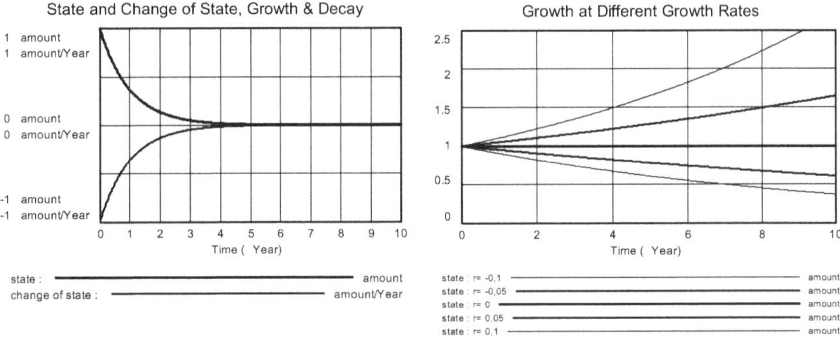

Figure Z103b: Negative feedback causes the state to decrease exponentially.
Figure Z103c: State development as function of time for different positive and negative growth rates.

If the sign is changed ($r = 1$), the process produces quick exponential growth. A typical characteristic of exponential growth is permanent growth of the absolute state increase, resulting in very high values of the state toward the end of the simulation period (here: 22'000 at $t = 10$).

If the change rate is positive, then the state grows exponentially with time towards infinity. If it is negative, then the state decreases exponentially with time towards zero. With exponential growth one has to keep in mind that the absolute change of state continues to increase exponentially even if the specific rate of change should remain constant.

Figure Z103c shows the state as function of time for different growth and decay rates in the range between -0.1 and +0.1. Many important dynamic processes, such as economic and population growth, occur in this range between 0 and ± 10% decay or growth per time unit.

The sign of the feedback obviously determines stability ($r < 0$) or instability ($r >$ 0) of the state change and system development. The rate of growth or decay is determined by the absolute value of the feedback $|r|$. The reciprocal $T = 1/|r|$ is referred to as the "time constant" of the system.

Technical note concerning the simulation: Feedback of this type is the reason that a (positive) state can never become negative even at a negative change rate since the change of state also approaches zero as the state disappears. By contrast, other formulations of the rate of change can easily lead to negative stocks (of cows, for example) and corresponding grave errors.

Exercises

1. Examine the temporal development of the *state* as a function of the feedback parameter RATE OF CHANGE r in the range from -1 to +1.
2. Examine the growth of a national economy at different growth rates (RATE OF CHANGE) between 0 and 10% per annum. To what multiple of the initial value will economic activities grow at different growth rates in this range within 100 years? Discuss how realistic these results are in consideration of corresponding resource and energy demands and the volume of consumption associated with it.
3. Derive the following approximate rule for exponential growth or decay from the mathematical formula for this process, and confirm the result by simulations:
"The number 70 divided by the change rate (in per cent per time unit) yields the doubling time (or: half-life period in a decay process) of a stock."
4. Consider a pollutant decomposing at a rate of 1% per year. How much time will it take until an initial amount of 100 tons is reduced to (a) one half, (b) 1/10, (c) 1/100 of the initial value? Confirm the result of the simulation with a mathematical calculation.
5. In the 1980's some scientists predicted an increase of primary energy consumption of the industrial nations of 3.5% per annum far into the 21st century, associated with a growth of electricity consumption of 7% per annum. Using an initial value of 100 per cent, what percent value is reached for these predictions after 10, 20, 50 and 100 years? Compare the result with the actual development. Can you think of reasons for such nonsensical forecasts?

Z104 Exponential delay

Simulation task

The dynamics of model Z103 "Exponential growth and decay" is entirely generated by endogenous processes, not by any external input. Growth or decay is driven by the state itself which correspondingly changes with time. Autonomous processes of this kind are quite common. They often confuse the observer, as an exogenous influence cannot be recognized – and, in fact, does not exist.

If systems with such internally produced "intrinsic" dynamics are simultaneously also subject to external inputs, their behavior becomes even more difficult to analyze and understand, since the effects of intrinsic dynamics and external inputs combine and cannot be traced to their individual source.

A simple first example is a system where the change of state is caused not only by the state itself (as in Z103) but also by a time-dependent external input (input function). An example is the time-dependent filling of a reservoir with simultaneous state-dependent loss of some of the stored quantity (e.g. by leakage, as in Z102D). Other examples: dynamics of soil water with a state-dependent percolation rate and time-dependent precipitation, time-dependent inputs of fertilizer and chemicals and their state-dependent uptake and absorption. In a mechanical system negative feedback in the system can cause a damping of its dynamics.

We now consider a model containing the negative feedback loop of model Z103 and in addition an external time-dependent input. The system dynamics now results from two different rates of change: (1) the exponential decay (exponential damping) resulting from the feedback of the state, and (2) a state-independent time-dependent input function. Thus the state is subject to a permanent exponential decay, but is constantly being "refilled" by the time-dependent inflow. Since the current state represents the time integral of previous changes, the losses correspond to a constant loss of the "history" of the system, while the current gains provided by the input function represent the present. The system state therefore represents mainly the most recent changes. The state follows the external input with a delay. This particular structure therefore functions as a delay, and is often used for this purpose in technical applications and in simulations.

Simulation model

The simulation model is shown in Figure Z104a and in the following simulation statements. It consists of model Z103 modified by the *input function*. This may consist of a PULSE SEQUENCE, STEP FUNCTION, RAMP FUNCTION, or SINE FUNCTION or a linear combination of these functions. The *rate of change* is composed of the *input function* and the feedback of the *state* multiplied by the RATE OF SELFCOUPLING.

Parameters and initial state
RATE OF SELFCOUPLING = -1 [1/Year]
INITIAL STATE = 0 [amount]
PULSE SEQUENCE = 0 [amount/Year]
STEP FUNCTION = 1 [amount/Year]
RAMP FUNCTION = 0 [amount/Year]

SINE FUNCTION = 0 [amount/Year]
FREQUENCY = 0.5 [1/Year]

Dynamics
input function = PULSE SEQUENCE *50 *PULSE TRAIN (1, 0.02, 1, 20) +STEP
      FUNCTION *STEP(1, 1) +RAMP FUNCTION *RAMP(1, 1, 5) +SINE FUNC-
      TION *SIN (2 *3.14159 *FREQUENCY *Time) [amount/Year]
rate of change = input function +RATE OF SELFCOUPLING *state [amount/Year]
state = INTEG (+rate of change, INITIAL STATE) [amount]

Simulation time parameters
INITIAL TIME = 0 [Year]
FINAL TIME = 10 [Year]
TIME STEP = 0.02 [Year]

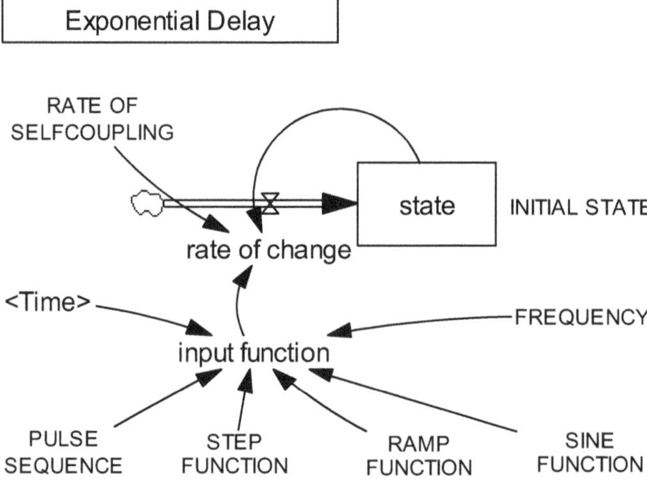

Figure Z104a: Simulation diagram for exponential delay.

Simulation results

Figure Z104b shows the response of the system to a step function (STEP = 1 at time t = 1) for an INITIAL STATE = 0. The *rate of change* initially corresponds to the *input function*, resulting in a linear increase of the *state*. Following the step, the input function remains constant. As the *state* increases, the negative feedback increases also, thus reducing the *rate of change*, which diminishes asymptotically towards zero. Eventually, the *input function* and the negative feedback of the *state* just balance, and a dynamic equilibrium results with *state* z^* = const. The transition to the equilibrium state is described by the exponential time function $z(t) = (u/r)\,(1 - e^{-rt})$. The *state* therefore approaches asymptotically the value of the constant input u divided by the feedback rate r.

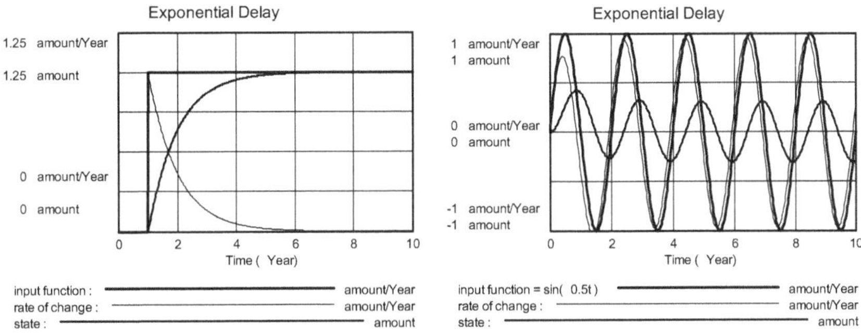

Figure Z104b: Response to a step function.
Figure Z104c: Delay of a periodic input function.

This delay effect becomes more obvious if we look at the development of the state as a function of a sinusoidal oscillation (with FREQUENCY = 0.5) as *input function* (Figure Z104c). After a brief adjustment phase, the *state* also oscillates with the frequency of the *input function*, although with a phase shift and smaller amplitude.

Because of this delay effect, the exponential delay (of first order, as shown here, but also of third order, as in model Z116) is often used to simulate delays. The critical parameter is the strength of the negative feedback (RATE OF SELFCOUPLING r). This parameter determines the dynamic equilibrium $z*$ of the state for constant input u. It corresponds to the condition $dz/dt = 0$, i.e. $rz* = u$. From this follows the equilibrium state: $z* = u/r$ (with $r > 0$). As r is increased (greater loss rate), the equilibrium state $z*$ becomes smaller. The inverse of the feedback parameter is the time constant of the system $T = (1/r)$. T is the most important system parameter. As r increases, T decreases, the system adjusts more quickly to the exogenous input $u(t)$, and the delay is shorter.

Exercises

1. Determine form and delay of the system response to different test functions (pulse, step, ramp, sine function with different frequencies), and to interesting combinations of these test functions.

2. Investigate the effect of the feedback strength (RATE OF SELFCOUPLING) on system response and delay of the input signal.

3. Develop analytical expressions for the phase shift and the amplitude ratio of a sinusoidal input as a function of the feedback parameter. Confirm the result by simulation.

Z105 Time-dependent growth

Simulation task

Structurally determined exponential growth may not be recognizable as exponential growth (or decay, depending on sign of feedback) if growth rates change by (externally determined) functions of time. Time dependence of specific growth rates is found, for example, in population development with time-dependent birth and death rates as function of changing medical care. Another example is plant growth as function of seasonally changing insolation or precipitation. Economic development also depends on time-variant investment decisions.

The basic structure of the model considered here corresponds to that for exponential growth or decay (as in models Z103 and Z104). The *change of state* again has two components: one (the feedback component) is proportional to the current *state*, the other depends on the *rate of change* provided as (externally determined) function of time. Since the *rate of change* changes with time, the *state* may increase as well as decrease in the course of time. Besides the strength of the *rate of change* its (possibly time-variant) sign is therefore an important determinant of system behavior.

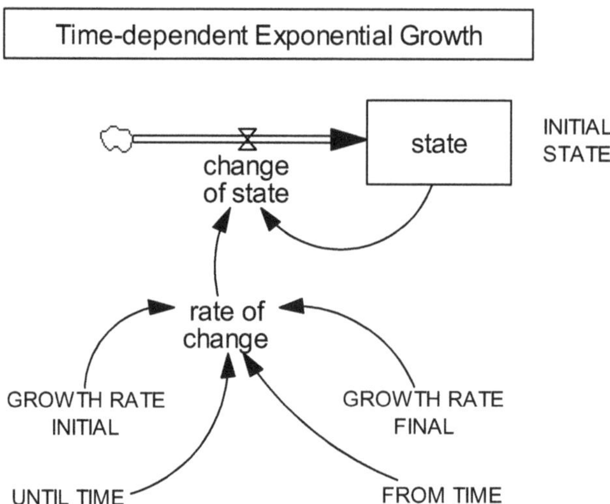

Figure Z105a: Simulation diagram for time-dependent change of state.

Simulation model

The model is documented in the simulation diagram of Figure Z105a and in the following program statements. Time dependence of the *rate of change* is modeled by a ramp function of time, as follows: Initially, and up to point UNTIL TIME the *rate of change* remains constant with value GROWTH RATE INITIAL. It then changes linearly with time until it reaches GROWTH RATE FINAL at time FROM TIME. From this time on *rate of change* remains constant at value GROWTH RATE FINAL. The four parameters of this time function can be adjusted as needed, allowing a large number of different cases to be investigated. The default values used here prescribe initial growth of 10% per year until year 2. Thereafter, *rate of change* decreases linearly to a value of 1% in year 8, and remains constant thereafter.

Parameters und initial states
GROWTH RATE INITIAL = 0.1 [1/Year]
UNTIL TIME = 2 [Year]
GROWTH RATE FINAL = 0.01 [1/Year]
FROM TIME = 8 [Year]
INITIAL STATE = 1 [amount]

Dynamics
rate of change = GROWTH RATE INITIAL +RAMP ((GROWTH RATE FINAL -
    GROWTH RATE INITIAL) /(FROM TIME -UNTIL TIME), UNTIL TIME, FROM
    TIME) [1/Year]
change of state = rate of change *state [amount/Year]
state = INTEG (+change of state, INITIAL STATE) [amount]

Simulation time parameters
INITIAL TIME = 0 [Year]
FINAL TIME = 10 [Year]
TIME STEP = 1 [Year]

Simulation results

The time function of the *rate of change* provided by the default settings results in an initially strong, then decreasing, and finally only weak growth of the *state*. Figure Z105b shows the corresponding results for *rate of change*, *change of state*, and *state*.

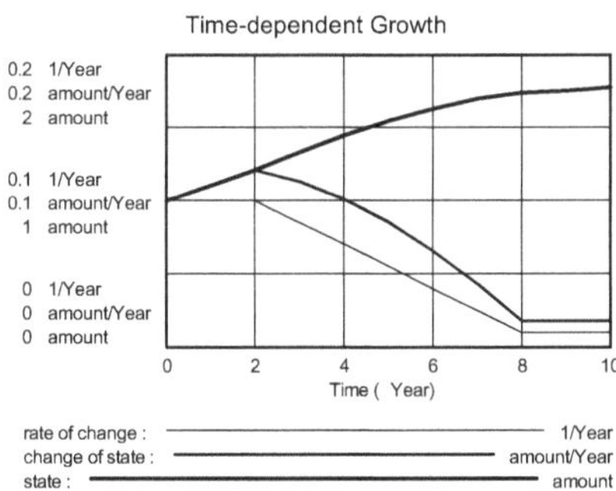

Figure Z105b: State development for linearly decreasing *rate of change*.

Exercises

1. Investigate the state development for different combinations of the scenario parameters (GROWTH RATE INITIAL, UNTIL TIME, GROWTH RATE FINAL, FROM TIME) for an INITIAL STATE = 1.

2. Introduce realistic values for the five parameters (year, rate of change, initial state) to simulate (approximately) developments such as the following: (a) population development in the USA, Great Britain (or other country) from 1950 to 2050; (b) economic growth in these countries between 1950 and 2010 (starting with a relative initial state of 100% in 1950); (c) development of global population between 1950 and 2050. Apply different scenario assumptions to determine the parameters for future conditions.

3. Replace the computation of the *rate of change* by a table function using historical values for the cases in the previous exercise. Compare the simulation results with the statistical data. Apply different (realistic) scenario assumptions to investigate the whole spectrum of possible future developments.

Z106 Simple population dynamics

Simulation task

The dynamics of many systems is determined by simultaneous exponential growth and decay processes, i.e. the presence of self-coupling feedback with positive and negative sign. Dynamic equilibrium results if at any time the gains equal the losses. *Example*: population growth as function of time-dependent changes in birth and death rates, in particular demographic transition, i.e. the stabilization of a population by birth rates becoming equal to death rates. This system structure applies generally to all processes with state-dependent change of state as function of growth and decay having time-dependent specific rates (e.g. pollutant concentration, CO_2 in the atmosphere). The same structure applies also to business dynamics, in particular to the calculation of orders, product stocks, accounts, and capital stocks, and to many other areas.

A structural characteristic of such systems are state-dependent inflows and out-flows having exogenously determined time-dependent inflow and outflow rates which are independent of each other. The absolute change of state is the difference between current inflows and outflows. The state can therefore decrease even at high inflow rates as long as losses are greater than gains. On the other hand, the state can increase even at small inflow rates as long as the loss rate is smaller than the gain rate. The time development of the system is therefore determined by the *net growth rate = gain rate – loss rate*. Control of one or both of these rates of change can therefore be used to obtain different developments of the state, including dynamic equilibrium.

If the state changes exponentially with constant gain and loss rates, the gains and losses per time unit remain at a fixed percentage of the state (e.g. +3% or -2% per year). The difference between gains and losses determines whether the state will grow exponentially or decay exponentially. If the gain rate equals the loss rate, the state remains at a constant value. This simple type of system structure can therefore produce three possible modes of behavior: growth, equilibrium, and decay.

Other modes of behavior, often differing greatly from these three modes, appear if the (relative, %) values of the gain and loss rates are functions of time. An example is population development as consequence of historically changing birth and death rates. In particular, the matching of gain and loss rates can result in stabilization at high or low levels. This process is of great significance for the stabilization of a popu-lation. Model Z106 "Population dynamics" allows investigation of development pos-sibilities for time-dependent gain and loss rates.

Development dynamics resulting from the competition of time-variable gain and loss rates is of fundamental importance for many areas of application. It will be studied here in terms of population dynamics, but it can be easily reformulated also for other applications.

Simulation model

We investigate two simulation models having essentially the same structure, but dif-fering in the formulation of birth and death rates. In the first model (Figure Z106Aa) birth rate and death rate remain constant, while they are functions of time in the sec-ond model (Figure Z106Ba). In this case the time-dependent rate of change (linear increase) is computed as in model Z105 "Time-dependent growth".

If the population doubles, the number of parents also doubles. If a constant number of births per couple is assumed, the number of births also doubles. Besides on population, the number of births also depends decisively on the birth rate, i.e. the number of births per year related to certain number of people (e.g. 1000). In Germany for example, the present birth rate is about 9/1000, i.e. 0.009.

Similar considerations apply to the computation of the number of deaths per year. Again, they are proportional to the number of people in the population. The number of deaths per year follows from multiplying the population by the death rate, which is also specified as number of deaths per 1000 people. In Germany, the death rate at present is about 12/1000, i.e. 0.012.

The population number after a certain period of time (e.g. 1 year) is computed by adding to the previous population number the births per year and subtracting the deaths per year. In the simulation diagram (Figure Z106A) this is indicated by the double arrow for the *births* pointing into the box for the state variable *population*, while the double arrow for the *deaths* points away from it.

The simulation diagram for population dynamics shows two feedback loops. The *birth* loop is a *positive* feedback loop: the larger the population, the more births; the more births, the larger the population etc. A steadily (exponentially) growing population would result if there were no deaths. But the gain is at least partially compensated by the *negative* feedback of the death loop: the larger the population, the larger the number of deaths, the smaller the population. Obviously there will be dynamic equilibrium if the number of births is exactly equal to the number of deaths. If the number of deaths outweighs the number of births, the population will shrink. This interplay of gain rate and loss rate can be found in many areas. It is typical of many ecological processes, where growth dominates initially, until finally an equilibrium state (climax) develops, where growth is just balanced by decay.

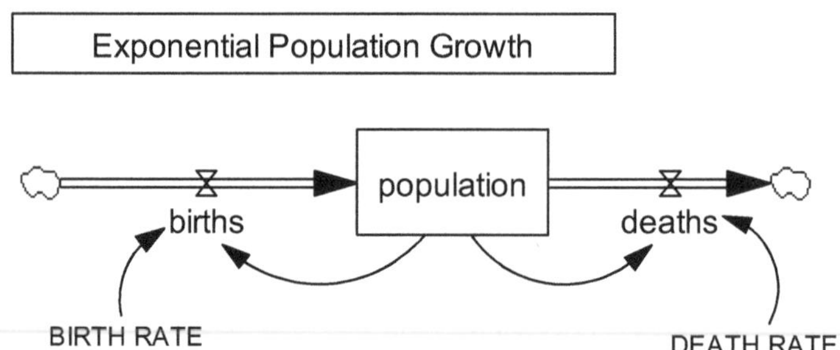

Figure Z106Aa: Simulation diagram for constant rates of change.

The following model equations apply to the computation of population dynamics for constant rates of change (Z106A):

Parameters
BIRTH RATE = 0.035 [1/Year]
DEATH RATE = 0.01 [1/Year]

Dynamics
population = INTEG (births -deaths,1e+006) [people]
births = BIRTH RATE *population [people/Year]
deaths = DEATH RATE *population [people/Year]

Simulation time parameters
INITIAL TIME = 0 [Year]
FINAL TIME = 20 [Year]
TIME STEP = 0.05 [Year]

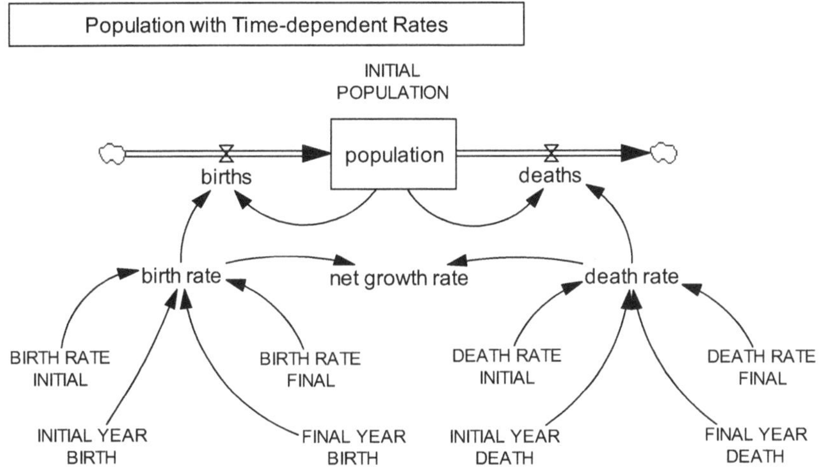

Figure Z106Ba: Simulation diagram for time-dependent rates of change.

If linear time functions are used for birth and death rates (model Z106B, Figure Z106Ba), the following model equations apply:

Parameters and initial state
INITIAL YEAR BIRTH = 2010 [Year]
BIRTH RATE INITIAL = 0.04 [1/Year]
FINAL YEAR BIRTH = 2060 [Year]
BIRTH RATE FINAL = 0.01 [1/Year]
INITIAL YEAR DEATH = 2010 [Year]
DEATH RATE INITIAL = 0.015 [1/Year]
FINAL YEAR DEATH = 2030 [Year]
DEATH RATE FINAL = 0.012 [1/Year]
INITIAL POPULATION = 1000 [people]

Dynamics
birth rate = BIRTH RATE INITIAL +RAMP ((BIRTH RATE FINAL -BIRTH RATE INI-
        TIAL) /(FINAL YEAR BIRTH -INITIAL YEAR BIRTH), INITIAL YEAR BIRTH,
        FINAL YEAR BIRTH) [1/Year]
death rate = DEATH RATE INITIAL +RAMP ((DEATH RATE FINAL -DEATH RATE
        INITIAL) /(FINAL YEAR DEATH -INITIAL YEAR DEATH), INITIAL YEAR
        DEATH, FINAL YEAR DEATH) [1/Year]

net growth rate = birth rate -death rate [1/Year]
births = birth rate *population [people/Year]
deaths = death rate *population [people/Year]
population = INTEG (+births -deaths, INITIAL POPULATION) [people]

Simulation time parameters
INITIAL TIME = 2000 [Year]
FINAL TIME = 2100 [Year]
TIME STEP = 0.1 [Year]

Obviously these models contain a number of simplifying assumptions. To test application validity in a concrete case, the underlying assumptions must be critically reviewed. In these simple models neither the consequences of limited carrying capacity nor effects of age structure, food supply, medical system, or random events are considered.

Simulation results

By way of example Figure Z106Ab shows die development over 20 years of an initial population of 1 million people at constant birth rate of 0.035 and death rate of 0.01 (typical for developing nations). With these parameters the population increases to about 165% of its initial value. After 20 years about 32'000 people are added to the population per year.

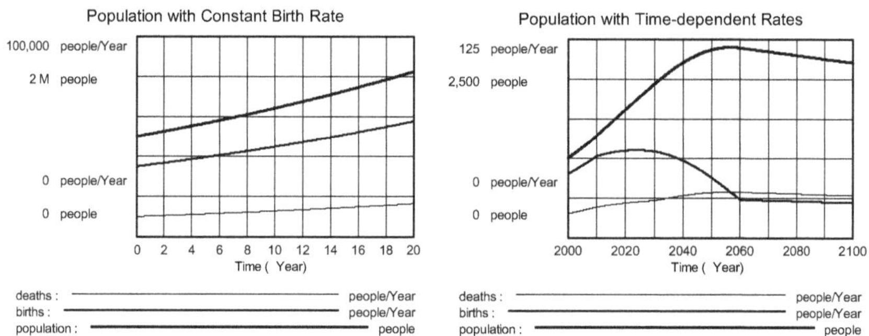

Figure Z106Ab: Population development at constant rates of change.
Figure Z106Bb: Population development at time variable rates of change..

Things look quite different if initially high birth and death rates are reduced in the course of time, in particular if the birth rate decreases faster than the death rate (Figure Z106Bb). In this example it was assumed that the birth rate decreases over 50 years from initially 0.04 to 0.01, while the death rate is reduced in 20 years from 0.015 to 0.012. In this case the initial population was 1000. (If this is understood as 1000 million, it corresponds roughly to the population of China or India.).

Note: For the case of population equilibrium, the death rate corresponds to the inverse of life expectancy, i.e. 0.012 corresponds to a life expectancy of 1/0.012 = 83 years. For this value of life expectancy a birth rate of 0.01 means that each woman (half of the population) has on average $0.01 \cdot 2 \cdot 83 = 1.66$ children. To stabilize a population, about 2.3 children per woman are necessary to replace deaths. The historical process of initially high birth rates adjusting to decreasing death rates in many countries is referred to as "demographic transition".

Exercises

1. Look up the current population numbers and birth and death rates for different industrial and developing nations and compute the population numbers that would result if birth and death rates would remain constant for the next 50 years.
2. Make meaningful assumptions for the future development of birth and death rates of these countries. Use linear approximation by defining initial and final values for these parameters. Enter the corresponding values into the simulation model. Compute the population development resulting from this scenario.
3. How can it be explained that the statistical death rates of developing countries are often quite low (e.g. about 0.01, which appears to indicate a life expectancy of $1/0.01 = 100$ years), although real life expectancy is actually quite low? (Assume that the death rate statistics are correct!).

References

More complex population models can be found in Bossel, H. 2007: *Systemzoo 3 Simulation Models – Economy, Society, and Development*. Books on Demand, Norderstedt.

Z107 Infection dynamics

Simulation task

The spreading of an infectious disease, the communication of a rumor or a good joke, the diffusion of innovations and other processes taking place in a population turn out on closer inspection to be saturation processes that can be computed by a corresponding simulation model. In the following, a model of this infection process will be developed.

If part of a population is infected by an infectious disease or some novel information, it can pass it on to the not yet infected part of the population when both groups come in contact. The number of these contacts and the probability of "infection" during contact will determine the spreading speed of the infection, i.e. the *rate of infection*. All we can say without further analysis is that the rate of infection will be low when either the number of infected or the number of yet to be infected is small, i.e. at begin and end of the epidemic. The rate of infection is highest if both the number of *infected* and the number of *not infected* is at a maximum, i.e. if exactly half of the (infectable) population has been infected.

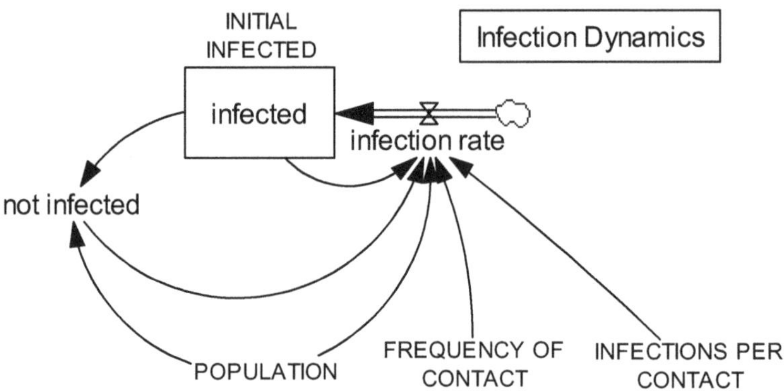

Figure Z107a: Simulation diagram of the infection process.

Simulation model

The simulation diagram for these relationships is shown in Figure Z107a. We are dealing with two stocks that are related to each other and therefore not independent of each other. Since the value of the total population remains constant, the number of *not infected* must obviously increase as the number of *infected* increases. For this reason only one of these quantities can be defined as state variable. We use the number of *infected*. The transition from the category "not infected" to that of "infected" is determined by the *infection rate*. It depends on the number of *infected*, the number of *not infected*, the FREQUENCY OF CONTACT, and the number of INFECTIONS PER CONTACT.

The corresponding complete simulation program is documented in the following. The parameters of the simulation can be changed before every simulation run.

Parameters
POPULATION = 4e+007 [people]
INITIAL INFECTED = 10 [people]
INFECTIONS PER CONTACT = 0.1 [1]
FREQUENCY OF CONTACT = 0.2 [1/Day]

Dynamics
infection rate = not infected *infected *FREQUENCY OF CONTACT *INFECTIONS
        PER CONTACT /POPULATION [people/Day]
infected = INTEG (infection rate, INITIAL INFECTED) [people]
not infected = POPULATION -infected [people]

Simulation time parameters
INITIAL TIME = 0 [Day]
FINAL TIME = 1500 [Day]
TIME STEP = 1 [Day]
SAVEPER = TIME STEP [Day]

Simulation results

Figure Z107b presents the results for the spreading of a disease, an innovation, or a joke in a population of 40 million people. The simulation time unit is [day]. Initially, only 10 persons are infected. Each person is assumed to meet a new person every 5 days (FREQUENCY OF CONTACT = 0.2 contacts per person per day). An infection is assumed to occur only during every tenth contact (INFECTIONS PER CONTACT = 0.1). The simulation period is 1500 days (i.e. somewhat more than 4 years). Remarkably, the infection spreads only very slowly during the first 600 days, and can hardly be recognized in the data. Thereafter, however, it progresses very rapidly, infecting almost the complete population in little more than a year. Different results are obtained, of course, if different values for the parameters are chosen.

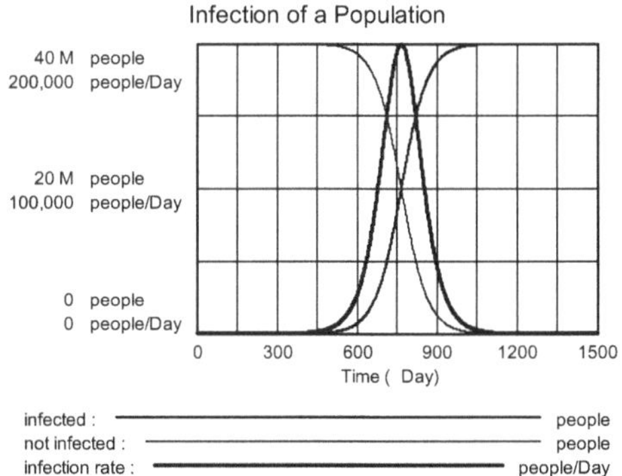

Figure Z107b: Simulation results for the infection process.

Exercises

1. Change the initial value, the time period of simulation, and the parameters to simulate the spreading of a rumor in a small town.

2. Assume that part of the population is IMMUNE to the disease by previous inoculation. Since these people cannot be infected, they must be counted with the *infected* (who can also no longer be infected). The value of INFECTIONS PER CONTACT must be reduced in proportion to the fraction IMMUNE/POPULATION. How does this effect the spreading of the epidemic? Simulate the effects of different inoculation strategies. What part of the population should at least be immunized by inoculation to prevent a spreading of the disease?

3. Expand the model to allow simulation of polarizing effects between two societal groups having high FREQUENCY OF CONTACT among their own group, but little with the other group.

References

K. Kalgraf: Yellow Fever Model. In M.R. Goodman: *Study Notes in System Dynamics*. Wright-Allen Press, Cambridge, Mass. 1974 (365-375).

Z108 Overloading a buffer

Simulation task

Permanent increases of state variables are not possible in reality – eventually, the capacity limit of their reservoirs is reached. In many cases the capacity limit makes itself felt long before it is reached, limiting the rate of increase in such a way that the capacity limit will not be exceeded. Several of the following models deal with the simulation of such processes.

However, it is also often the case that the current state level has no influence on the inflow. In this case the reservoir can fill up to its capacity limit unhindered and will then overflow. The overflow rate is usually significantly greater than the normal outflow. As the inflow is again reduced to its normal level, the high rate of overflow will speed up the reestablishment of the normal dynamics of inflow and outflow of the reservoir.

Systems with this kind of overflow dynamics are relatively common: overflow of a reservoir dam after snow melt or after a rainstorm; rapid surface run-off of rain water when the soil has become water-logged and cannot take up any more precipitation; overload of body organs (e.g. iodine uptake of thyroid gland), where substances that can no longer be absorbed are eliminated at a high rate if a critical limit is exceeded. There are many other processes in everyday life that follow these dynamics of "overloading a buffer".

Simulation model

To describe the complete overflow dynamics, the overflow process must be modeled in addition to the normal flow process. The corresponding simulation diagram is shown in Figure Z108a; the model equations are listed in the following.

In this model it is assumed that normal *outflow* is proportional to current *buffer level* and NORMAL OUTFLOW RATE. For low *buffer level* the *outflow* is therefore reduced. However, if the *buffer level* exceeds the BUFFER CAPACITY, the excess causes *overflow* at an OVERFLOW RATE considerably higher than NORMAL OUTFLOW RATE. This process continues until *buffer level* is again reduced to below BUFFER CAPACITY, and normal *outflow* is reestablished.

To simulate shock loads, a strong pulse-shaped additional *inflow* of amount PULSE HEIGHT is added to the NORMAL INFLOW, starting at time PULSE BEGIN and lasting for a time period PULSE LENGTH.

Parameters and initial state
BUFFER CAPACITY = 1 [amount]
INITIAL BUFFER LEVEL = 0.3 [amount]
NORMAL INFLOW = 0.25 [amount/Day]
NORMAL OUTFLOW RATE = 0.5 [1/Day]
OVERFLOW RATE = 10 [1/Day]
PULSE BEGIN = 5 [Day]
PULSE LENGTH = 0.2 [Day]
PULSE HEIGHT = 10 [amount/Day]

Dynamics
inflow = NORMAL INFLOW +PULSE HEIGHT *PULSE (PULSE BEGIN, PULSE
        LENGTH) [amount/Day]
outflow = NORMAL OUTFLOW RATE *buffer level [amount/Day]
overflow = IF THEN ELSE (buffer level > BUFFER CAPACITY, OVERFLOW RATE
        *(buffer level -BUFFER CAPACITY) , 0) [amount/Day]
total outflow = outflow +overflow [amount/Day]
buffer level = INTEG (+inflow −outflow -overflow, INITIAL BUFFER LEVEL) [amount]

Simulation time parameters
INITIAL TIME = 0 [Day]
FINAL TIME = 20 [Day]
TIME STEP = 0.02 [Day]

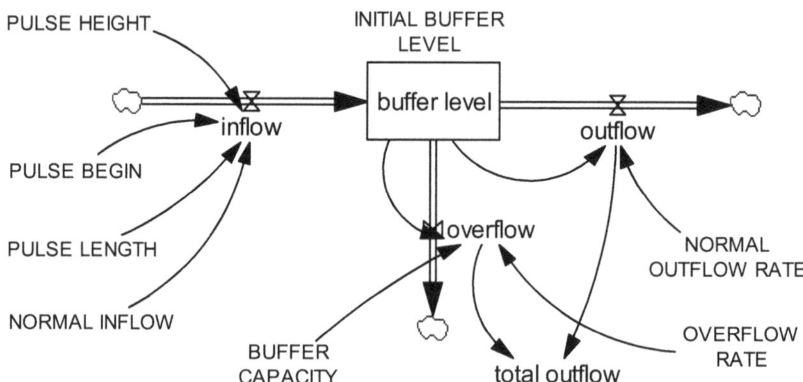

Figure Z108a: Simulation diagram overloading a buffer.

Simulation results

The simulation results for the default parameters of the model are shown in Figures
Z108b and Z108c. In the second diagram the development between days 4 and 6 is
graphed with a stretched time axis, showing the overflow dynamics more clearly.

For the NORMAL INFLOW = 0.25 assumed here and the normal *outflow* of (NOR-
MAL FLOW RATE · *buffer level*) = 0.5 · *buffer level* an equilibrium state is established at
a *buffer level* of 0.5. Since the INITIAL BUFFER LEVEL is 0.3 in this example, the *buffer
level* increases initially, reaching the equilibrium value around day 5. At this time the
additional inflow pulse of PULSE HEIGHT = 10 occurs, which lasts for 0.2 [day] and
causes the buffer to overflow after a short time. This starts the excessive overflow,
which continues even after the pulse inflow has stopped until the *buffer level* has
dropped below the BUFFER CAPACITY ($k = 1$). Normal *outflow* resumes thereafter.
Until the equilibrium value (at *time* > 10 days) has been reached, the *buffer level* drops
continuously (to 0.5).

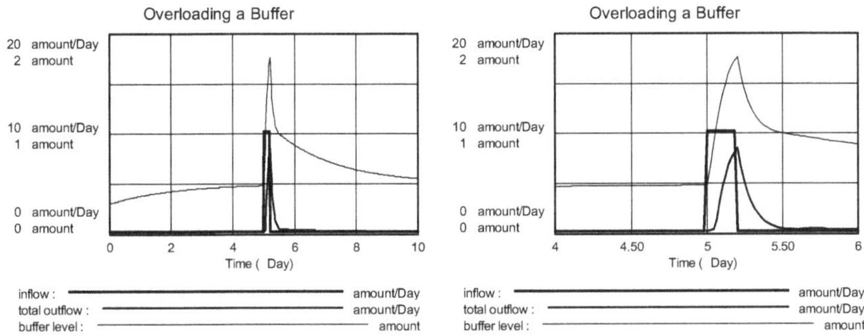

Figure Z108b: Overload and overflow for a sudden shock load.
Figure Z108c: The same process shown with stretched time axis.

The behavior of the system depends strongly on BUFFER CAPACITY k and NORMAL OUTFLOW RATE r: As long as k and/or r are large, overload will not easily occur. For the condition $k \cdot r > u(t)$ (i.e. if $k \cdot r$ is always greater than the *inflow u*), overload cannot occur. Flood catastrophes can therefore be avoided if reservoirs or retention basins are large enough to take up excessive inflows. If applied to soil water dynamics, this result implies that water-logging is not to be expected if either the soil water capacity k is high, the percolation rate r is high, and/or the precipitation input $u(t)$ remains small. These conditions do not apply, for example, if the soil layer is thin and consists of clay, or after a heavy downpour.

The significance of the BUFFER CAPACITY becomes obvious when it is raised high enough (to 2.5) to accommodate the inflow pulse without overflowing (Figure Z108d). Under otherwise identical conditions a "flood" does not now occur; the normal *outflow* increases only slightly. This relative constancy of outflow – caused by large underground aquifers – is characteristic of many springs where the outflow hardly changes with time despite large fluctuations of rainfall. (Cf. model Z301 "Regional watershed" in Bossel 2007 *System Zoo 2*).

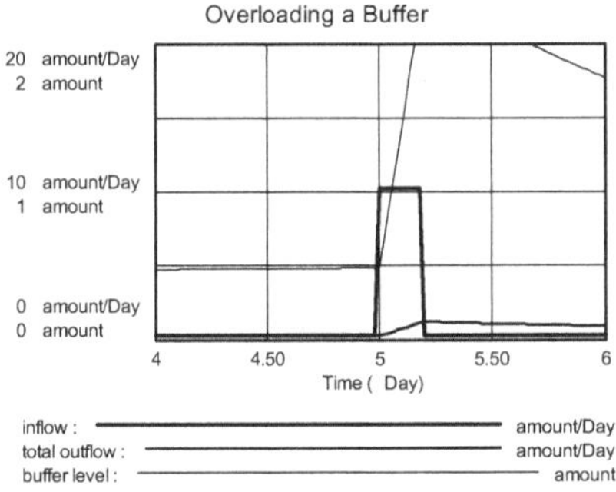

Figure Z108d: There is no "flood event" if the reservoir capacity is large enough.

Exercises

1. Investigate the influence of the buffer capacity for avoiding flood peaks and over-flow of the buffer.

2. Choose model parameters (including correct choice of units!) realistically reflecting the conditions for a lake or water reservoir in your region, its normal inflows and possible inflow pulses caused by snowmelt or heavy rainfall. Are the simulation results for overflow and possible flooding events realistic? If not, try to find the reasons for strong deviations from observations.

3. Investigate repeated heavy rainfall episodes by using a pulse train sequence. Investigate in particular the influence of pulse period and pulse area (pulse height · pulse width) on the dynamics and possible flood peaks.

Z109 Logistic growth with constant harvest

Simulation task

In models Z105 "Time-dependent growth" and Z106 "Simple population dynamics" the increase or loss rates are provided as constant or time-dependent quantities. In this case the rates (not the absolute values of the increases or losses) were therefore independent of the respective stock. However, feedback of the stock size to the growth rate is to be expected if the stock reaches the limits of the environmental carrying capacity (overpopulation) or if, on the other hand, the stock drops below a minimum size no longer allowing reproduction (extinction of rare animals). In both cases the rate of increase or decrease is a function also of the stock size itself.

Growth with saturation because of absorption or carrying capacity limitation is found in many processes. Examples are the growth of organisms (animals and plants) up to a certain, genetically determined size, adaptation of herbivore populations to available pasture, market saturation of a new product etc. An inverse, but structurally similar process can arise if the loss rate increases (or the growth rate is strongly reduced) as the stock drops below a certain minimum level. Finally, erosion processes where the erosion rate steadily increases with progressive erosion are also related to this process.

The logistic growth process is of fundamental importance. It is found in almost all fields: ecology, economy, technology etc. It stands out by the fact that initially, while the stock is still small, growth is almost exponential. As the stock approaches its capacity limit, negative feedback becomes increasingly effective, finally forcing the stock to remain at the capacity limit. The development dynamics of the stock has a typical S-shaped (sigmoid) form. If the stock is harvested at constant harvest rate, the equilibrium state remains below the capacity limit if the rate is small enough. If however the harvest rate exceeds a critical limit, then the collapse of the complete stock becomes unavoidable.

Simulation model and simulation results

We first develop the simulation model for the logistic growth dynamics of a stock neglecting losses (e.g. by harvesting), see the simulation diagram of Figure Z109Aa and the following model equations.

The GROWTH RATE of an exponential growth process with positive self-coupling of the state is multiplied by the *remaining capacity factor* = (1 − *state*/CAPACITY). If the *state* approaches the CAPACITY, this factor approaches zero and the *change of state* is also reduced to zero. Thereafter, the *state* remains at its CAPACITY value.

Figure Z109Ab shows the time development of *change of state* and *state* for the default parameter setting. The INITIAL STATE is defined as minute value (0.01) in the integration relation (INTEG) for the state. (An initial state of "0" would not produce any result.) The *state* describes an S-curve whose initial increase is characterized by the GROWTH RATE. While the modifying factor (*state* * *remaining capacity factor*) reduces the (specific) rate of growth from its initial value GROWTH RATE to the final value zero, the resulting *change of state* increases from zero to a maximum at the inflection point of the S-curve for the state, before it again drops to zero.

The system has still another interesting variant: If the parameter CAPACITY is a negative quantity (it then no longer has the meaning of a capacity limit or load-capacity), then the *state* grows to infinity in finite time (finite escape time).

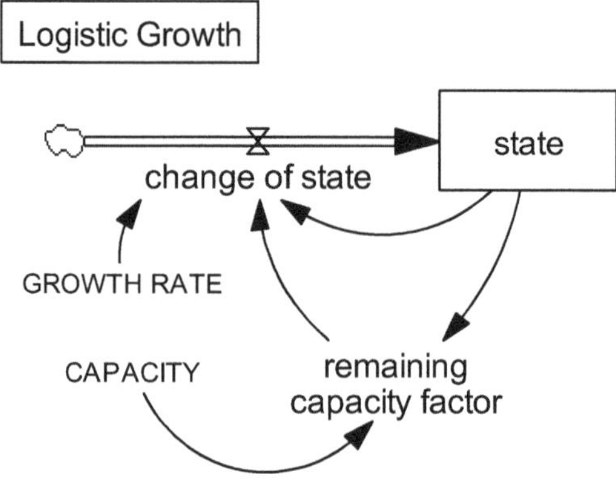

Figure Z109Aa: Simulation diagram for logistical growth.

Parameters
CAPACITY = 1 [amount]
GROWTH RATE = 1 [1/Year]

Dynamics
remaining capacity factor = 1-(state/CAPACITY) [1]
change of state = GROWTH RATE *state *remaining capacity factor [amount/Year]
state = INTEG (change of state, 0.01) [amount]

Simulation time parameters
INITIAL TIME = 0 [Year]
FINAL TIME = 20 [Year]
TIME STEP = 0.05 [Year]

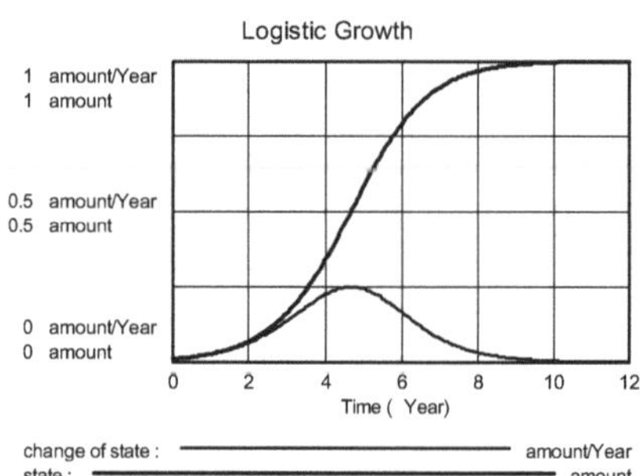

Figure Z109Ab: Logistic growth of a stock.

The logistic growth function describes many events in nature quite well, such as the growth of organisms or populations to their genetic or ecological capacity limit. However, what happens if such a population is harvested, that is if e.g. a fish population is permanently harvested by a particular amount per year?

A corresponding model addition to simulate harvesting has been introduced in the simulation diagram of Figure Z109Ba and in the following model equations. In this case harvesting occurs at a constant HARVEST RATE. Figure Z109Bb shows simulation results for different values of the HARVEST RATE. At a low HARVEST RATE an equilibrium state of the *stock* establishes itself below the CARRYING CAPACITY k. If a critical HARVEST RATE e is exceeded, the *stock* breaks down completely in short time (obvious in this case for $e = 0.27$). This erosion process is a common phenomenon: If the load exceeds a critical limit, a system cannot be saved any more from complete collapse.

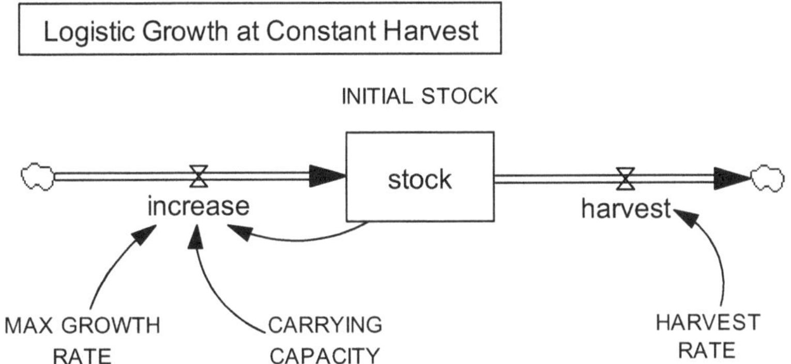

Figure Z109Ba: Simulation diagram for stock-independent harvest.

Parameters and initial state
INITIAL STOCK = 1 [amount]
CARRYING CAPACITY = 1 [amount]
MAX GROWTH RATE = 1 [1/Month]
HARVEST RATE = 0 [amount/Month]

Dynamics
stock = INTEG (+increase -harvest, INITIAL STOCK) [amount]
harvest = HARVEST RATE [amount /Month]
increase = MAX GROWTH RATE *stock *(1 −stock /CARRYING CAPACITY)
	[amount/Month]

Simulation time parameters
INITIAL TIME = 0 [Month]
FINAL TIME = 20 [Month]
TIME STEP = 0.05 [Month]

For constant HARVEST RATE e a stable equilibrium develops as long as $e < rz\,(1 - z/k)$. The stability limit is defined by the maximum permitted HARVEST RATE $e = rk/4$. Even if this value is only minimally exceeded, the system breaks down. Thus the maximum yield lies exactly at the stability limit. The system has two equilibrium points of which, however, only one is stable.

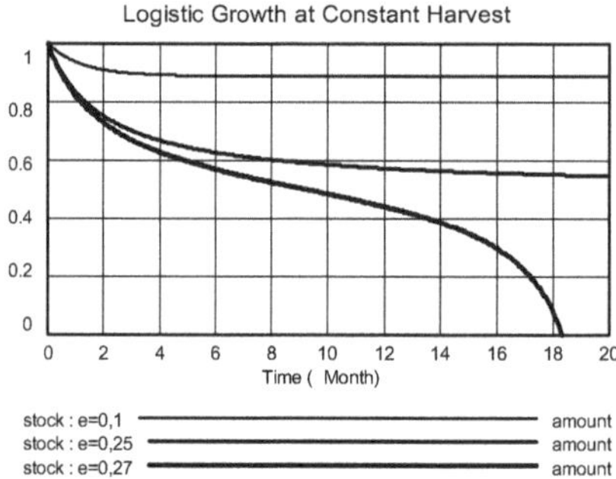

Figure Z109Bb: Stock independent harvest can lead to the collapse.

Exercises

1. Examine the development of the stock and its (net) growth rate starting with different initial conditions, at HARVEST RATES of $0.1 < e < 0.9$.

2. Examine the growth curves of different populations with different initial states, saturation values and rates of increase. Use appropriate time units in the model.

3. Use the model for studying the finite escape time with capacity (or CARRYING CAPACITY) $k = -1$ (harvest rate $e = 0$). Examine the behavior for different values of rate of *increase* or MAX GROWTH RATE r $(0.1 < r < 0.5)$. Store results of different runs and draw them in a common diagram.

4. The logistic growth process can be understood also as a process with simultaneous positive and negative feedback, where initially positive feedback provides exponential growth, and negative feedback produces growth equilibrium as the capacity limit is approached. Develop the corresponding simulation model (simulation diagram and model equations) and show that it is identical with model Z109A.

5. Using the model, simulate the growth of a wheat field from the beginning of the vegetation period (April 15th) up to harvest (August 15th) (use [day] as time unit). The *biomass* (organic dry matter) is initially 150 [kg/hectare] (seed), growing to 12 000 [kg/hectare] (*grain + straw + plant remains* = CAPACITY) at HARVEST TIME. Choose the *growth rate* such that the logistic growth curve reaches its saturation (= CAPACITY) shortly before HARVEST TIME.

6. Expand the model to compute the following case (use [year] as time unit): Let the carrying capacity limit for a whale population be 100 000. Let the minimum value (erosion limit) of the population be 5000. If the whale stock drops below this value, reproduction stops (birth rate = 0) since partners do not find each other any more (insert corresponding IF condition). Let the normal (specific) birth rate = 0.1, the (specific) death rate = 0.05 (complete the model by including losses through deaths). Assume the stock is reduced to 20 000 by over-harvesting. Compute the further development of the stock assuming harvest quotas of 20, 10, 5, 2.5 and 0 per cent per year.

References

Richter, O. 1985: *Simulation des Verhaltens ökologischer Systeme*. VCH Weinheim.

France, J., Thornley, J. H. M. 1984: *Mathematical Models in Agriculture*. Butterworths, London (Ch. 5).

Luenberger, D. G. 1979: *Introduction to Dynamic Systems – Theory, Models, and Applications*. John Wiley, New York (317-319).

Kormondy, E. J. 1976: *Concepts of Ecology*. Prentice-Hall, Englewood Cliffs, N.J., 1976 (75-132).

Ehrlich, B. R., Ehrlich, H. H., Holdren, P. 1977: *Ecoscience: Population, Resources, Environment*. Freeman, San Francisco (97-122, 181-246).

Busch, K.F., Uhlmann, D., Weise, G. (eds.) 1983: *Ingenieurökologie*. G. Fischer, Jena (53-63).

Colinvau, P. 1993: *Ecology 2*. John Wiley, New York (141-149).

Z110 Logistic growth with stock-dependent harvest

Simulation task

The simulations with model Z109 "Logistic growth with constant harvest" have shown that a population under harvest can quickly collapse if losses from the harvest (e.g. from fishing) cannot be compensated by regeneration of the population. Particularly worrisome is the fact that maximum yield is obtained exactly at the collapse limit. If harvesting aims for maximum yield, eventual collapse is practically inevitable.

Sustainable harvesting of populations, e.g. sustainable fishery, must therefore be managed such that (1) collapse of the harvested population is avoided, and (2) a high sustainable yield is realized. This can only be achieved if harvesting is guided by reference to the current population size. This means that the harvest rate must be reduced if stocks decline. In the following, the logistic model is modified to include this condition.

The model describes very generally the logistic growth of a state with a state-dependent loss rate. Important applications are fishing or hunting, for example: While a fish population can easily collapse if fish locator technology is used (and most fish can be easily located and caught), this is not possible if locator technology is not applied. In that case the catch rate drops inevitably as fish population density is reduced by catches, and the population cannot collapse (cf. Bossel 1994: 196-212, 135-141, 265-273). Another example comes from economics: To avoid ruining the economic existence of enterprises, taxes should be a function only of the current active production rate.

Simulation model

The model is documented in the simulation diagram of Figure Z110a and in the following model equations. A *stock z* with logistic growth (parameterized by MAX GROWTH RATE r and CARRYING CAPACITY k) constantly loses a certain fraction by *harvest* which is proportional to the existing *stock* and a given constant HARVEST RATE e. The ratio of HARVEST RATE / MAX GROWTH RATE $= e/r$ can have an arbitrary value. With this particular system structure collapse of the system is impossible: As the *stock* decreases, so the *harvest* is reduced by feedback from the *stock*, and the *stock* can not completely disappear.

Parameters and initial state
INITIAL STOCK = 0.01 [amount]
MAX GROWTH RATE = 1 [1/Year]
HARVEST RATE = 0.2 [1/Year]
CARRYING CAPACITY = 1 [amount]

Dynamics
increase = MAX GROWTH RATE *stock *(1 −stock /CARRYING CAPACITY) [amount/Year]
harvest = HARVEST RATE *stock [amount/Year]
net growth = increase -harvest [amount/Year]
stock = INTEG (+increase -harvest, INITIAL STOCK) [amount]

Simulation time parameters
INITIAL TIME = 0 [Year]
FINAL TIME = 20 [Year]
TIME STEP = 0.02 [Year]

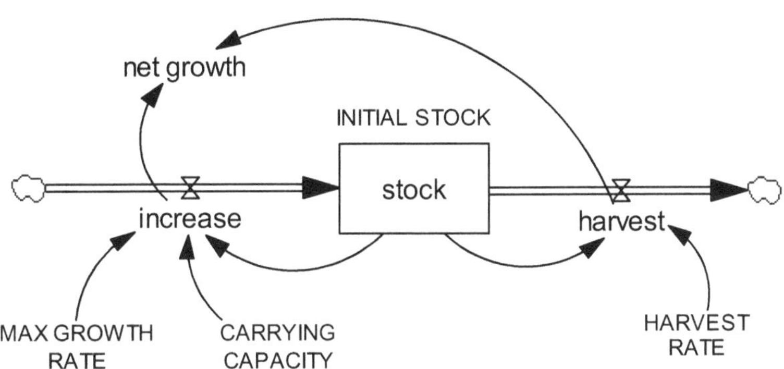

Figure Z110a: Simulation diagram for stock-dependent harvest.

Simulation results

Depending on the specific HARVEST RATE, the system state (*stock*) approaches a corresponding equilibrium value. Figure Z110b shows the result for the default parameter settings. The HARVEST RATE of $e = 0.2$ leads to an equilibrium value of *stock* $z^* = 0.8$ and to a corresponding sustainable *harvest* of 0.16. The equilibrium state is given by $z^* = k\,(1 - e/r)$ for $e = r$.

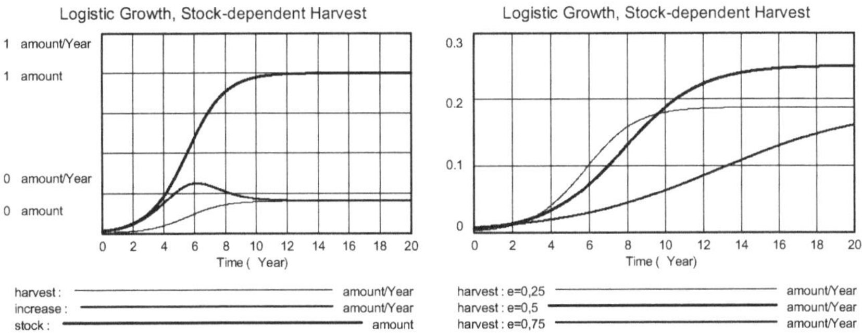

Figure Z110b: If harvest is proportional to current stock, the system cannot collapse.
Figure Z110c: Maximum sustainable yield is achieved at moderate harvest rate.

The question for which HARVEST RATE e the highest possible *harvest* E_{max} can be achieved is of practical importance. This is the case for $e/r = 0.5$. It follows from this that $e\,z^* = E_{max} = e\,k\,(1 - e/r) = e\,k\,(1 - 0.5) = e\,k\,/2$. This can be verified by a simulation (with $k = 1$) as shown in Figure Z110c. Lower values for *harvest* are found for $e = 0.25$ and $e = 0.75$. *Harvest* is maximum for $e = 0.5$, where it reaches a value of $E_{max} = 0.25$.

For stock-dependent harvest the specific HARVEST RATE e is not critical for the existence of the stock, but it determines the equilibrium level and the *harvest*. For maximum yield the HARVEST RATE e should be equal to one half of the MAX GROWTH RATE r. The maximum *harvest* is equal to that for stock-independent constant harvest (as in model Z109). However, in the present case this harvesting constellation is stable, and the stock is not in danger of collapse. For $e = 0$ and $k < 0$ we again obtain finite "escape time" (as in model Z109A).

Exercises

1. Derive the relations given above for the equilibrium value *stock* z^* and the maximum achievable *harvest* E_{max}.
2. Compare the behavior, the achievable *harvest* and the stability characteristics with model Z109 "Logistic growth with constant harvest".
3. Investigate the development of a *stock* for different initial conditions, MAX GROWTH RATES r and HARVEST RATES e.
4. Use the model with realistic parameter assumptions describing e.g. a sustainable fishery in a large lake or sustainable hunting for deer in wooded terrain. Determine for these cases the required HARVEST RATE for maximum harvest yield and the corresponding *harvest*.

References

Cf. model Z109 "Logistic growth with constant harvest".

Z111 Density-dependent growth (Michaelis-Menten)

Simulation task

Saturation processes are found in many applications, but they cannot always be described correctly by a logistic formulation. The reaction rate V of chemical processes and particularly of catalytic reactions is primarily dependent on the concentration S of the substrate in addition to environmental factors such as temperature, pH value etc. The process can be represented by so-called Michaelis-Menten kinetics:

$$\frac{V}{V_{\max}} = \frac{S}{c+S}$$

Here c is the so-called half saturation constant, i.e. the substrate concentration at $V_{\max}/2$. For the state $S = c$ the resulting saturation is 50%. The Michaelis Menten term (right hand side) is zero if $S = 0$, 0.5 if $S = c$, and 1 if $S \to \infty$.

Simulation model

In model Z111 (see the simulation diagram of Figure Z111a and the following model equations) a saturation term of the Michaelis-Menten form $z/(z + c)$ is used to obtain the S-shaped saturation behavior of the *stock z*. The remaining structure resembles that of the logistic growth process with an exponential growth loop $r{\cdot}z$ and a *harvest e·z* (or corresponding deaths) dependent on population density. (r = MAX GROWTH RATE, z = *stock*, e = HARVEST RATE)

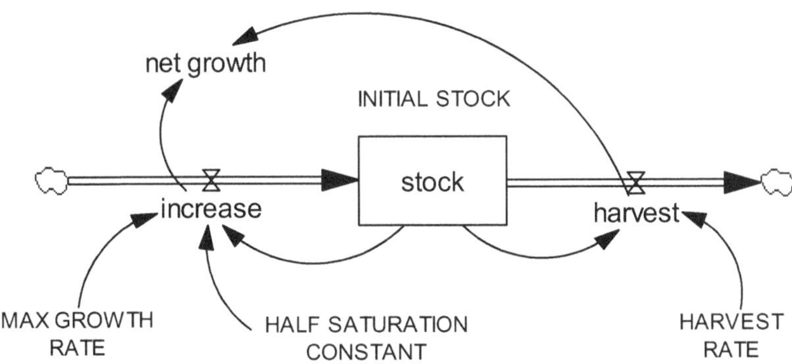

Figure Z111a: Simulation diagram for Michaelis-Menten kinetics.

Parameters and initial state
INITIAL STOCK = 0.02 [amount]
MAX GROWTH RATE = 1 [1/Day]
HARVEST RATE = 0.5 [1/Day]
HALF SATURATION CONSTANT = 1 [amount]

Dynamics
increase = MAX GROWTH RATE *stock *(1-stock /(HALF SATURATION CONSTANT
     +stock)) [amount/Day]
harvest = HARVEST RATE *stock [amount/Day]
net growth = increase -harvest [amount/Day]
stock = INTEG (+increase-harvest, INITIAL STOCK) [amount]

Simulation time parameters
INITIAL TIME = 0 [Day]
FINAL TIME = 50 [Day]
TIME STEP = 0.02 [Day]

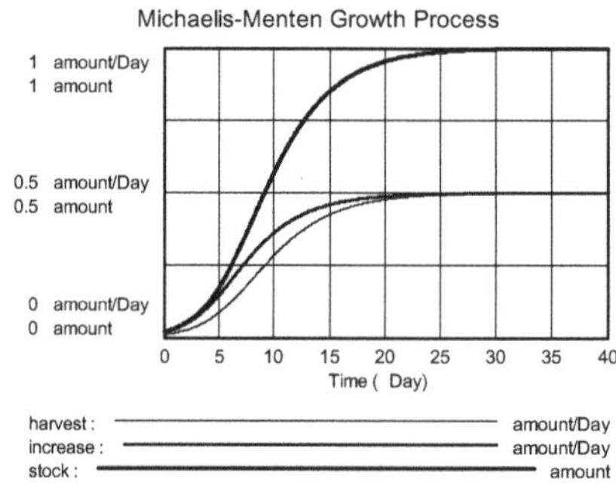

Figure Z111b: Growth of a stock with Michaelis-Menten kinetics.

Simulation results

Figure Z111b shows the simulation result for the default parameter settings listed
above. Since at first the *increase* is larger than the *harvest*, the stock grows in an S-
shaped curve beginning with INITIAL STOCK and ending at the saturation value. The
following condition applies for the equilibrium state z^* (saturation value):

$$z^* = (\frac{cr}{e}) \cdot (1 - \frac{e}{r})$$

Obviously no finite saturation value can be reached if the system is not harvested (i.e.
$e = 0$). This also follows from simulation with HARVEST RATE = 0 (Figure Z111c). In
this case a constant increase of the stock results as *time* $\rightarrow \infty$. This is an important
difference compared with the logistic model, which always achieves a finite equilibrium state, even if there is no harvesting. In Michaelis-Menten kinetics, saturation can
only occur if $e > 0$.

If the HARVEST RATE e is greater than the MAX GROWTH RATE r, the *stock* decays in an (approximately) exponential pattern (Figure 111d). As long as e remains <
r, the *stock* cannot collapse. The equilibrium state depends more critically on the ratio
r/e than is the case in the logistic system.

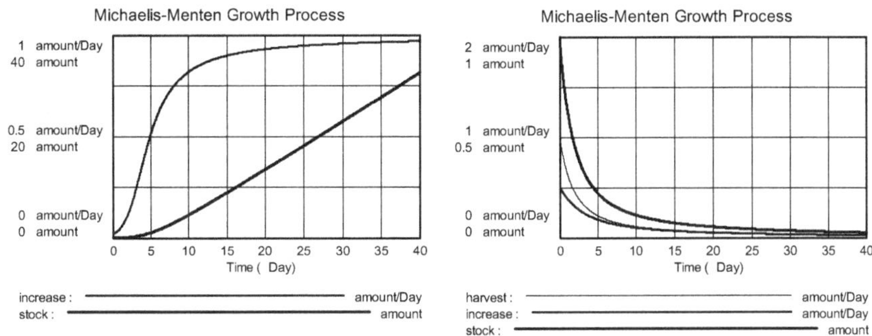

Figure Z111c: Unrestrained growth for the case of no harvesting.
Figure Z111d: Decay of a stock (INITIAL STOCK = 1) at high HARVEST RATE = 1.

Exercises

1. Examine the behavior of the system for different MAX GROWTH RATE r at a constant HARVEST RATE $e > 0$. Store the results of every run and then draw the different resulting curves as a function of *time* in a common diagram.
2. Examine in the same way the behavior for different HARVEST RATES $e > 0$ at constant MAX GROWTH RATE r.
3. Study the behavior for different values of the HALF SATURATION CONSTANT c.
4. Examine the behavior for different values of INITIAL STOCK.
5. Look up the description of a simple reaction with Michaelis-Menten kinetics and the necessary data in a chemistry textbook and develop a simulation model for the process. Compare your simulation result with the data given in the source, and find and correct possible mistakes.

References

Richter, O. 1985: *Simulation des Verhaltens ökologischer Systeme*. VCH Weinheim (178-187)
Straŝkraba, M., Gnauck, A. 1983: *Aquatische Ökosysteme – Modellierung und Simulation*. Gustav Fischer, Jena (100-102).

Z112 Double integration and exponential delay

Simulation task

Many physical processes incorporate a double integration of state quantities over time. In mechanics, for example, the velocity of a mass arises from the integration of its acceleration over time; further integration of the velocity produces the location of the mass as function of time. An equivalent description is the time integration of the acceleration force $m(dv/dt)$ to obtain the momentum mv; further time integration yields the kinetic energy $mv^2/2$. A structurally equivalent double integration arises from the summation of the voltage drops in an electric circuit consisting of capacitance, resistance and inductance.

The state at each integrator develops completely differently if each state is fed back to its own input by self-coupling, thus effecting its rate of change. This was evident in model Z103 "Exponential growth and decay" and was used in model Z104 "Exponential delay" to achieve a delay effect (by negative self-coupling, i.e. damping). For double integration a similar effect has to be expected if self-coupling of the two state variables is present.

Undamped integration (no self-coupling) corresponds to the analytical integration rules. However, if the two integrators have damping (i.e. negative self-coupling feedback), then an input signal which is modified and delayed in the first integrator is modified and delayed again in the second integrator. The result is qualitatively and quantitatively considerably different from undamped integration.

Simulation model

The model is documented in the simulation diagram in Figure Z112a and in the following model equations. The *change of state 1* is determined from the *input function*, the *state 1*, and the SELFCOUPLING FACTOR 1, and is integrated to yield *state 1*. In a similar way, *state 1* is used to determine the *change of state 2* and *state 2*. In both integrations the respective *state* is fed back to its *change of state* by self-coupling with the respective SELFCOUPLING FACTOR to compute the *change of state*. The model employs the usual test functions, which are activated (or deactivated, value = 0) by choosing corresponding parameters PULSE SEQUENCE, STEP FUNCTION, RAMP FUNCTION, AND SINE FUNCTION (as shown in model Z101).

Parameters and initial states
FREQUENCY = 0.1 [1/Month]
PULSE SEQUENCE = 0 [amount/Month]
RAMP FUNCTION = 0 [amount/Month]
SINE FUNCTION = 0 [amount/Month]
STEP FUNCTION = 1 [amount/Month]
SELFCOUPLING FACTOR 1 = 1 [1/Month]
SELFCOUPLING FACTOR 2 = 1 [1/Month]
INITIAL STATE 1 = 0 [amount]
INITIAL STATE 2 = 0 [amount*Month]

Dynamics
input function = PULSE SEQUENCE *50 *PULSE TRAIN (1, 0.02, 1, 20) +STEP
    FUNCTION *STEP (1, 1) +RAMP FUNCTION *RAMP (1, 1, 5) +SINE FUNC-
    TION *SIN (2 *3.14159 *FREQUENCY *Time) [amount/Month]
change of state 1 = input function -SELFCOUPLING FACTOR 1 *state 1 [a-
    mount/Month]
change of state 2 = state 1 -SELFCOUPLING FACTOR 2 *state 2 [amount]
state 1 = INTEG (change of state 1, INITIAL STATE 1) [amount]
state 2 = INTEG (change of state 2, INITIAL STATE 2) [Month*amount]

Simulation time parameters
INITIAL TIME = 0 [Month]
FINAL TIME = 10 [Month]
TIME STEP = 0.02 [Month]

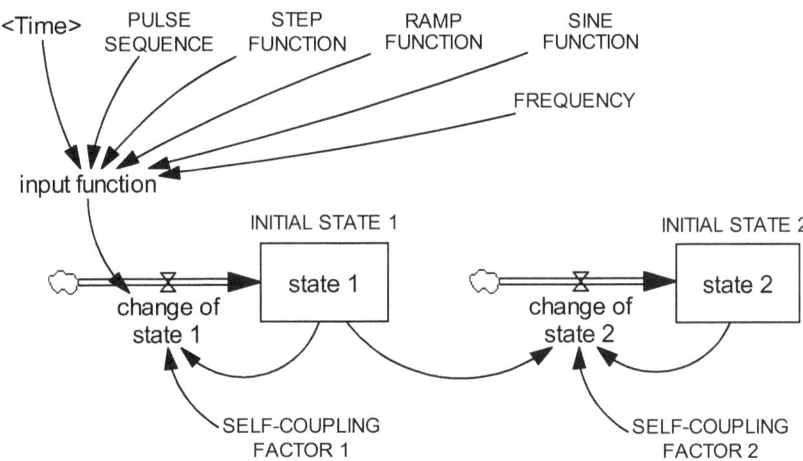

Figure Z112a: Simulation diagram for double integration and exponential delay.

Simulation results

In a first test we shall check whether the simulation model carries out the double integration correctly. For this test the two SELFCOUPLING FACTORS must be set to zero, INITIAL STATE 1 must be set to the correct initial value (here = 0), and INITIAL STATE 2 must be set to zero. A STEP FUNCTION = 1 at *time* = 1 is selected as *input function* (cf. default setting of parameters). The result is shown in Figure Z112b. Applying integration rules, the integration result for *time* = 10 is found to be *state 1* = 9 and *state 2* = $9^2/2 = 81/2 = 40.5$. This result is confirmed by the simulation (Figure Z112b).

Next, consider the effect of damping. For this purpose the two SELFCOUPLING FACTORS are set to the same nonzero value ($a = 1$ in the default setting). If we now use the same *input function* (STEP FUNCTION = 1 at *time* = 1) as in the previous case, completely different behavior results (compare Figure Z112c with Z112b). With self-coupling, *state 1* and *state 2* do not grow indefinitely, but adjust with some delay to the input signal. Initially, *state 1* increases linearly, *state 2* quadratically. Figure Z112d shows *change of state 1* and *change of state 2* for this case. The changes of state disappear when *state 1* and *state 2* have reached their equilibrium values.

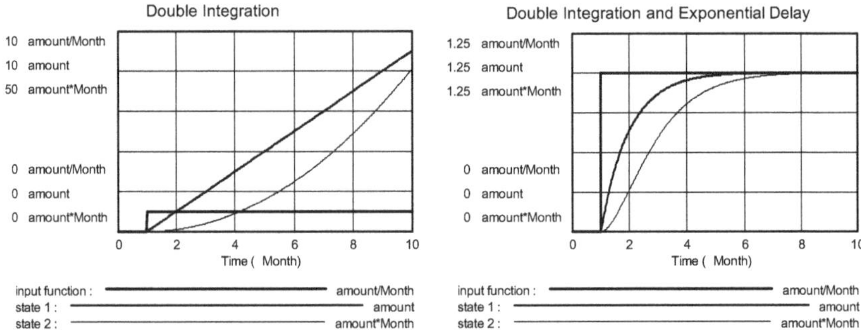

Figure Z112b: Double integration of a step function.
Figure Z112c: Delay effect of double integration.

The damping factor *a* determines the equilibrium values for the two state variables *state 1* and *state 2* for constant input u = const. The equilibrium state is given by $z_1^* = u/a$, $z_2^* = u/a^2$. This should be checked by adjusting the two SELFCOUPLING FACTORS, for example a = 0.5, 2, 3 etc. Their values also determine the delay effect. For strong negative feedback (large absolute value of *a*) the delay effect is small, since the time constant of the system $T = 1/a$ is now small. The delay period over both integrators follows from the sum of both time constants: $T_D = 1/a + 1/a = 2/a$.

The delay effect becomes even more obvious if a sine function (SINE FUNCTION = 1, FREQUENCY = 0.5) is chosen as an input function (Figure Z112e). The sinusoidal input oscillation causes both state variables to oscillate with the same frequency, but with different phase shift and amplitude.

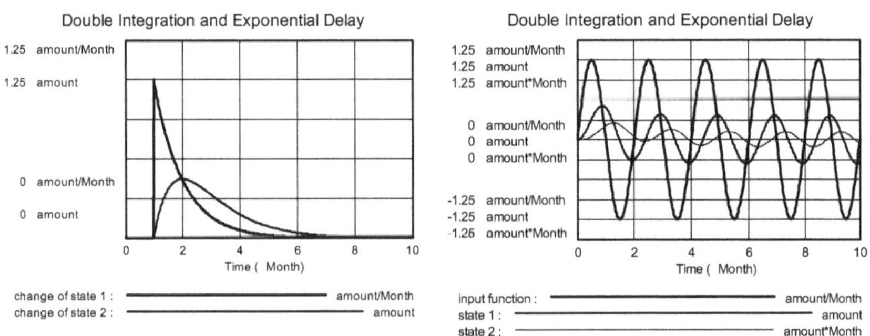

Figure Z112d: Changes of state for step function as input function.
Figure Z112e: Delay effect for a periodical input function.

Exercises

1. Choose different initial conditions and different input functions and compute the response of the model system. Check the result using the relevant mathematical integration formulae. How can possible small differences be explained?

2. Confirm (by simulation) the results from the formula (above) for the equilibrium values for constant input function for different (positive) values of a.

3. Construct your own test function $u(t)$ of some interesting shape as a sum of the test functions pulse sequence, step, ramp and/or sine oscillation. Integrate these once and twice numerically ($a = 0$).

4. Investigate how a step function is changed and delayed for small and large damping factors a. Calculate analytically the initial slopes and equilibrium values of *state 1* and *state 2* (see above) and check whether they correspond to the simulation results.

5. What conditions must be satisfied by an input function to ensure that the system returns to the original state (e.g. $z_1 = z_2 = 0$) after passing of the input function?

6. What phase shift will a sinusoidal input experience in the first and second integrator (*state 1*, *state 2*)? (Calculate analytically and check by simulation!)

7. How will the integrator state change with time as function of the initial values of the two integrators (z_1 and/or $z_2 > 0$), if the input function $u = 0$?

8. What distance is covered in 100 seconds by a body being accelerated with a constant acceleration of 1 [m/sec^2] (about 1/10 of earth gravity)? Check the simulation result by analytical computation. What explains the difference? How can it be (almost) avoided?

References

Bossel, H. 1994: *Modeling and Simulation*. A K Peters, Wellesley, MA (162-171).
Bossel, H. 2004: *Systeme, Dynamik, Simulation – Modellbildung, Analyse und Simulation komplexer Systeme*. Books on Demand, Norderstedt (178-179).

Z113 Transition from one state to another

Simulation task

People, organisms, products and other things often stay in a certain state or at a certain place for a certain time before they change to another state or place, and later perhaps even to another. *Examples*: Small children turn into students, then into working adults and parents, then into pensioners; butterfly eggs turn into caterpillars, then into pupae, then into butterflies; consumer goods are produced at one place, then warehoused, then transported to the dealer, then sold to customers; as rainwater, fertilizer or chemicals seep in the ground, they are held for shorter or longer periods by physically different soil layers before reaching the next layer.

Each of these stages is coupled to the previous and the following stages by transition processes. The current stock at every stage is a result of the initial state, the gains by inflows and the losses by outflows during the time interval under investigation. Again, the individual states have to be determined by integration of their inflows and outflows over time. However, for this process it is important to note that all stocks keep their identity, i.e. the dimension of all stocks must be identical. The transition processes must be defined accordingly. This process therefore differs fundamentally from integration in series as in the case of model Z112 "Double integration and exponential delay", where the dimension (unit of measurement) changes with every further integration by the factor [time unit].

Simulation model

The model is documented in the simulation diagram of Figure Z113a and in the following model equations. The model describes the gradual transition of the contents of one reservoir to the next one. The *input 1* to *state 1* corresponds to the given *input function*. The *state 1* accumulates in the first reservoir according to *input 1* and *transition* to the second reservoir. There *state 2* builds up according to gains by *transition* and losses by *output 2*.

Both *transition* and *output 2* are defined by the *state* of the preceding reservoir (*state 1*, *state 2*) and the corresponding specific flow rates (TRANSITION RATE, OUTPUT RATE 2) of dimension [1/time unit]. As a consequence, all three flows (*input 1*, *transition*, *output 2*) are proportional to the preceding state and also have correct dimensions [state unit / time unit]. The losses of *state 1* resulting from the *transition* are equal to the gains of *state 2* from this *transition*.

In this model the *transition* is proportional to the *state* of the giver reservoir. This is not mandatory: transition rates may in principle also be determined by the receiver reservoir or by a combination of giver and receiver stocks as in the case of the predator-prey system (cf. models Z401 through Z404 in *System Zoo 2*).

The mean residence time T_1 in *state 1* can be determined from the specific TRANSITION RATE r_1 ; the same is true for OUTPUT RATE and *state 2*. If for example r_1 = 0.1, then 1/10 of the stock leaves *state 1* per time unit. The mean residence time is therefore the inverse of the specific transition rate, here: 10 [time units].

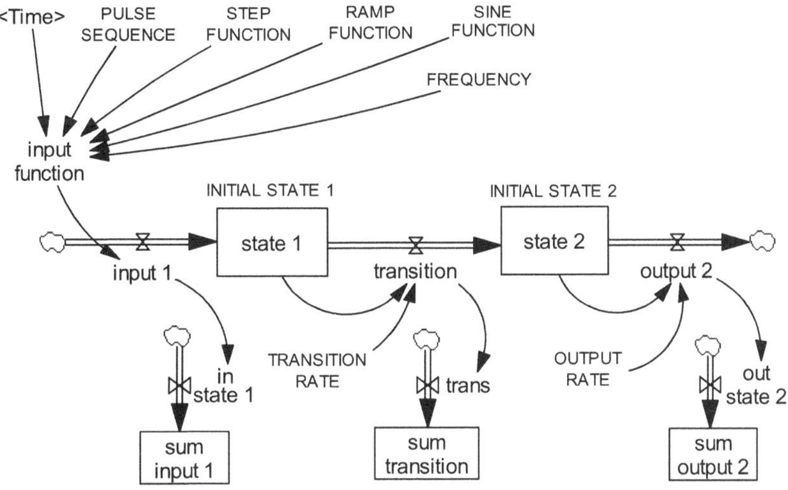

Figure Z113a: Simulation diagram for transition between two states.

Parameters and initial states
INITIAL STATE 1 = 0 [amount]
INITIAL STATE 2 = 0 [amount]
TRANSITION RATE = 0.5 [1/Second]
OUTPUT RATE = 0.25 [1/Second]
PULSE SEQUENCE = 1 [amount/Second]
STEP FUNCTION = 0 [amount/Second]
RAMP FUNCTION = 0 [amount/Second]
SINE FUNCTION = 0 [amount/Second]
FREQUENCY = 0.1 [1/Second]

Dynamics
input function = PULSE SEQUENCE *1 *PULSE TRAIN (1, 2, 50, 50) +STEP FUNC-
TION *STEP(1, 1) +RAMP FUNCTION *RAMP(1, 1, 5) +SINE FUNCTION *SIN
(2 *3.14159 *FREQUENCY *Time) [amount/Second]
input 1 = input function [amount/Second]
transition = TRANSITION RATE *state 1 [amount/Second]
output 2 = OUTPUT RATE *state 2 [amount/Second]
state 1 = INTEG (input 1 -transition, INITIAL STATE 1) [amount]
state 2 = INTEG (transition -output 2, INITIAL STATE 2) [amount]

Throughput sums
in state 1 = input 1 [amount/Second]
sum input 1 = INTEG (Ein 1, 0) [amount]
trans = transition [amount/Second]
sum transition = INTEG (Über, 0) [amount]
out state 2 = output 2 [amount/Second]
sum output 2 = INTEG (Aus 2, 0) [amount]

Simulation time parameters
INITIAL TIME = 0 [Second]
FINAL TIME = 20 [Second]
TIME STEP = 0.02 [Second]

Simulation results

Simulation results for the default parameter values are shown in Figures Z113b through Z113d. At *time* = 1 the input pulse of height =1 and duration =2 (predefined in the equation for *input function*) jumps to its value "1" and remains there until *time* = 3, when it returns to zero (pulse repetition is programmed in PULSE TRAIN for time = 50, which is outside the time span considered here). Corresponding to *input 1* the *state 1* increases rapidly, while already losses occur from *transition* = TRANSITION RATE · *state 1* (Figure Z113c). The *transition* now leads to an increase of *state 2*, which decreases again as its inflow by *transition* disappears and outflow by *output 2* reduces *state 2*. Altogether, it turns out that the "wave" caused by the initial pulse moves with time delay through the different states and is strongly smoothed in this process. Again, a delay effect arises for time-variable input $u(t)$ in this system at every stage of the transition process.

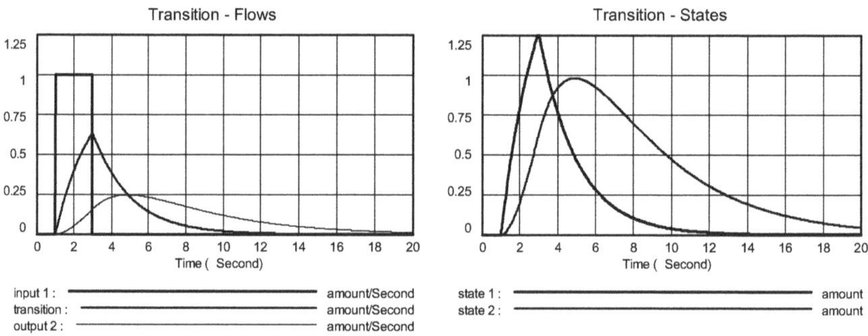

Figure Z113b: Changes (flows) in the transition process for a pulse input.
Figure Z113c: States in the transition process for a pulse input.

Since in this case no losses occur in the transition process, and the complete outflow of *state 1* enters *state 2* again, the time-sums over the simulation period of each of the flows of *input 1*, *transition*, and *output 2* must be identical. This can be checked by integrating the respective flows over time. Figure Z113d shows that the time-sums of the flows at these three locations actually approach the final value = 2 toward the end of the simulation period, corresponding to the pulse area = 2 from pulse height = 1 and pulse width = 2.

The behavior of the first integrator *state 1* (z_1) corresponds exactly to that of the exponential delay with feedback factor r_1 (TRANSITION RATE r_1). An equilibrium value of $z_1{}^* = u/r_1$ therefore establishes itself there also. The second integrator *state 2* (z_2) also has the basic structure of the exponential delay. The equilibrium value of $z_1{}^* = u/r_1$ found at the first integrator leads (as a constant input to the second integrator) to an equilibrium value of $z_2{}^* = u/(r_1 r_2)$ (where OUTPUT RATE = r_2).

The loss rates of the reservoirs (r_1, r_2) determine in this system also the change rate and with that the time constant, the delay, and the equilibrium state.

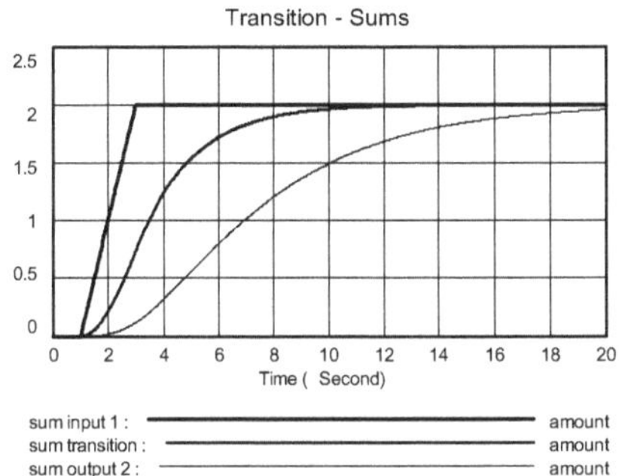

Figure Z113d: In the transition process the time-sums of all flows are identical.

Exercises

1. Examine the behavior of the state variables *state 1* and *state 2* for different values of TRANSITION RATE r_1 and OUTPUT RATE r_2.

2. Determine how different test functions (particularly sine functions) are "processed" by this system.

3. Why is this model not suitable for the simulation of transportation delays (conveyor belts, freight transport, production operations)? How would such delays have to be simulated?

Z114 Linear oscillator of second order

Simulation task

Oscillations are of great importance in our world. Systems which can produce oscillations are found in many areas: spring-mass-damper system; pendulum; electric oscillating circuit; interaction of market, capital and production; musical instruments. Oscillations can be damped (as in the case of vehicle suspensions), or can tend to amplification leading to destruction (as in the case of suspension bridges in a strong wind, or flutter of aircraft wings), or can even be used for measurement of time because of their regularity (pendulum clock, quartz watch).

Generally, oscillations can always be expected if a feedback loop connects more than at least two state variables, i.e. if a perturbation is fed back after a time delay to its origin, and the process therefore repeats periodically as a result of this delay with a certain frequency defined by the system parameters. However, whether a system of this type will oscillate also depends on sign and strength of the couplings in the system. A system with the basic structure of the oscillator can therefore exhibit periodical and aperiodic, damped and undamped behavior depending on its parameters.

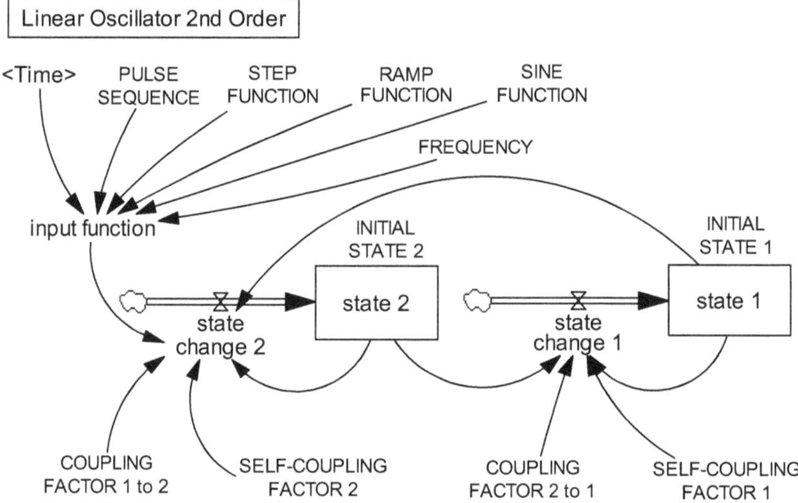

Figure Z114a: Simulation diagram of a system of second order capable of oscillation.

Simulation model

The simulation model is documented in Figure Z114a and in the following model equations. In the linear oscillator of second order (with two state variables) considered here a feedback loop connects the two state variables *state 1* and *state 2*. The feedback connections have weights c (COUPLING FACTOR 1 to 2) and b (COUPLING FACTOR 2 to 1). The motion can be damped or amplified depending on the signs of the self-coupling connections with parameters a (SELFCOUPLING FACTOR 1) and d (SELFCOUPLING FACTOR 2). Apart from oscillations other (aperiodic) motions are also possible.

To be able to examine a great variety of behaviors of the oscillator, several test functions are provided in the simulation model (PULSE SEQUENCE, STEP FUNCTION, RAMP FUNCTION, SINE FUNCTION with specification of FREQUENCY). They affect *state 2* by way of the *input function*.

Parameters and initial states
INITIAL STATE 1 = 0 [amount*Second]
INITIAL STATE 2 = 1 [amount]
a = SELFCOUPLING FACTOR 1 = 0 [1/Second]
d = SELFCOUPLING FACTOR 2 = -1 [1/Second]
c = COUPLING FACTOR 1 to 2 = -1 [1/(Second*Second)]
b = COUPLING FACTOR 2 to 1 = 1 [1]
PULSE SEQUENCE = 0 [amount/Second]
STEP FUNCTION = 0 [amount/Second]
RAMP FUNCTION = 0 [amount/Second]
SINE FUNCTION = 0 [amount/Second]
FREQUENCY = 0.1 [1/Second]

Dynamics
input function = PULSE SEQUENCE *50 *PULSE TRAIN (1, 0.02, 100, 100) +STEP
        FUNCTION *STEP(1, 1) +RAMP FUNCTION *RAMP (1, 1, 5) +SINE FUNC-
        TION *SIN (2 *3.14159 *FREQUENCY *Time) [amount/Second]
rate of change 1 = SELFCOUPLING FACTOR 1 *state 1 +COUPLING FACTOR 2 to
        1 *state 2 [amount]
rate of change 2 = COUPLING FACTOR 1 to 2 *state 1 +SELFCOUPLING FACTOR
        2 *state 2 +input function [amount /Second]
state 1 = INTEG (+rate of change 1, INITIAL STATE 1) [amount]
state 2 = INTEG (+rate of change 2, INITIAL STATE 2) [Sec*amount]

Simulation time parameters
INITIAL TIME = 0 [Second]
FINAL TIME = 20 [Second]
TIME STEP = 0.02 [Second]

Simulation results

The default parameter setting does not provide any input function, but defines INITIAL STATES $z_2 = 1$, and $z_1 = 0$. In this case the dynamics is not caused by any external influence. Figure Z114b shows the result for this parameter setting ($a = 0$, $b = 1$, $c = -1$, $d = -1$): The system oscillates strongly damped.

The behavior of the system is strongly dependent on the choice of the coupling parameters. In the following investigations $a = 0$ and $b = 1$ remain constant, and behavior is studied as function of COUPLING FACTOR 1 to 2 ($= c$) and SELFCOUPLING FACTOR 2 ($= d$). Figure Z114c shows the time diagram of four different runs, Figure Z114d shows the state trajectories (state diagram) for the same runs. With $c < 0$ and $d < 0$ damped oscillations arise. For $c < 0$ and $d = 0$ an undamped oscillation develops. From the time diagram Z114c the frequency f of this undamped natural oscillation with an oscillation period of approximately 6.3 seconds can be determined, i.e. $f = 1/6.3 = 0.159$. The analytically determined natural frequency is $\omega_0 = (|bc|)^{1/2}/(2\pi) = 1/(2\pi) = 0.1592$.

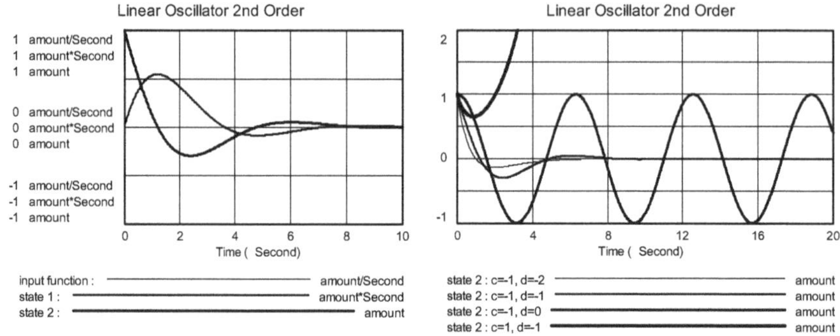

Figure Z114b: Damped oscillation (default parameter setting).
Figure Z114c: Parameter change causes change in behavior.

Note that these oscillations arise without the system being activated by a periodical input function. The oscillations are entirely caused by the given structural connections of the system components. This also becomes obvious if the sign of the COUPLING FACTOR 1 to 2 is reversed ($c > 0$). Both figures show that in this case the system produces steadily increasing state values in unstable aperiodic motion.

Figures Z114e and Z114f show the response of the system to excitation by a sine function (FREQUENCY f = 0.5) as *input function*. The forcing frequency is higher than the natural frequency of the undamped case (d = 0) in Figure Z114c. Initially, the system attempts to oscillate with its natural frequency, but it is quickly forced to adapt to the faster forcing frequency.

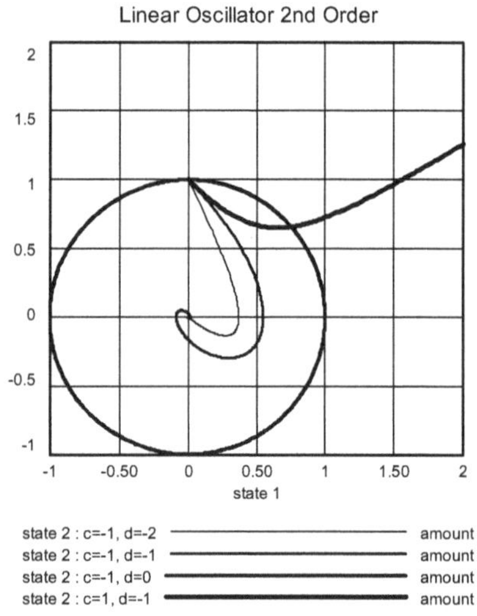

Figure Z114d: State trajectories for the cases of Figure Z114c.

Periodic excitation can cause oscillations to amplify even in a damped system – possibly leading to its destruction. Figures Z114g and Z114h show rapid amplification for $d = -0.01$ and $f = 0.16$ (close to the natural frequency).

Background

The system is defined by the following system of differential equations:

$$\frac{dz_1}{dt} = a\,z_1 + b\,z_2$$

$$\frac{dz_2}{dt} = c\,z_1 + d\,z_2$$

The system is linear and therefore has only one equilibrium point. In unforced motion (*input function u* = 0), its equilibrium coordinates ($z_1{}^* = 0$ and $z_2{}^* = 0$) are independent of the initial conditions of the state variables.

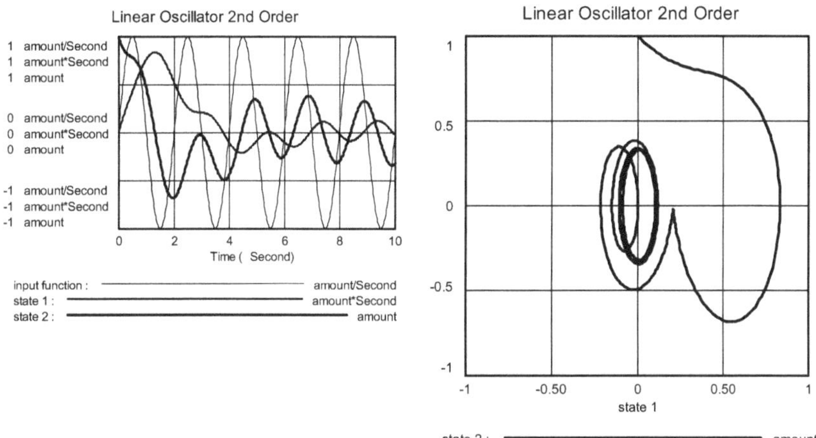

Figure Z114e: Response to sinusoidal excitation (FREQUENCY = 0.5).
Figure Z114f: State trajectory for the sinusoidal excitation of Figure Z114e.

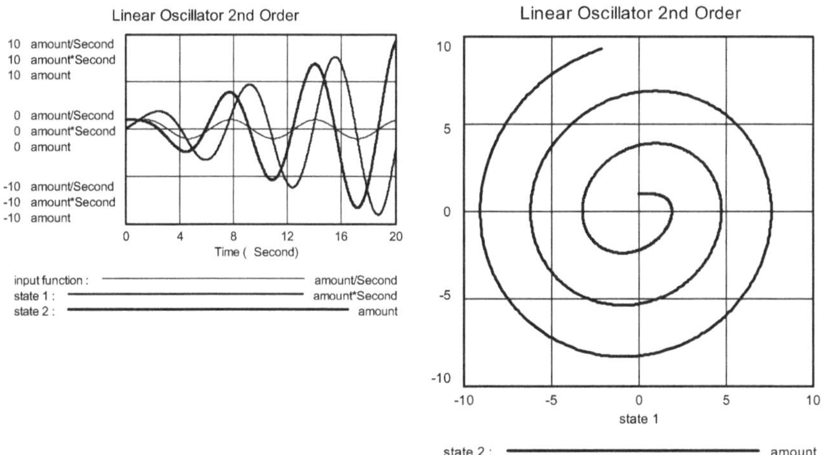

Figure Z114g: Oscillation with increasing amplitude resulting from periodic excitation near the natural frequency.
Figure Z114h: State trajectory for periodic forcing near the natural frequency of Figure Z114g.

Depending on the eigenvalues λ_1, λ_2 of the system matrix, fundamentally different possibilities of behavior arise which can be stable or unstable: source and sink, vortex (center) and focus, saddle, node, line source, line sink. By a corresponding choice of parameters a, b, c, d of the system matrix each of these behavior types can be generated with the model. A small variation of one or several system parameters may cause a fundamental change of behavior since the eigenvalues of the system may then shift to other regions of the complex plane. Eigenvalues located in the right hand complex plane (real part of eigenvalue > 0) always cause instability, eigenvalues in the imaginary region (imaginary parts of eigenvalue $\neq 0$) always produce oscillations.

For the system parameters of the reference run ($a = 0$, $b = 1$, $c = -1$, $d = -1$) the eigenvalues are $\lambda_{1,2} = -0.5 \pm 0.866i$. The two eigenvalues are conjugate complex; the imaginary part of both eigenvalues indicates oscillation, the real part with a negative sign means damping. (The theoretical background is treated in greater detail in Bossel 1994: 116-119, 388-432; and Bossel 2007: 91-93, 124-140).

Exercises

1. At which value of SELFCOUPLING FACTOR 2 $= d$ for the reference case (default parameter settings) will oscillating motion be replaced by aperiodic motion?
2. What is the consequence of a sign change of COUPLING FACTOR 1 to 2 $= c$ for system behavior? Try to generate all possible fundamental behaviors by changing the system matrix accordingly. *Note*: Use $a = 0$ and values for b, c, d of either +1 or -1.
3. Using the test function parameters SINE FUNCTION, FREQUENCY, and PULSE SEQUENCE and the default parameters, investigate the response (in particular: resonance) for forced oscillation.
4. Show that a pulse with pulse area $= 1$ (e.g. pulse height $= 50$, pulse width $= 0.02$, see default values) causes the same behavior of the system as an INITIAL STATE 2 $= z_2$ $= 1$. How can this be explained?
5. For INITIAL STATES $= 0$ and nonexistent damping ($a = 0$, $d = 0$), determine the natural frequency as function of (negative) COUPLING FACTOR 1 to 2 ($= c$) by forcing the system with a single pulse. Document the results in a time diagram. Compare them with the theoretical result for the natural frequency.
6. Force the system (using the default parameter settings) with sinusoidal oscillations in the frequency domain $0 < f < 10$. Determine the (constant) amplitude of the oscillation eventually developing for each forcing frequency. Plot this amplitude as a function of the frequency (frequency response). Where does this curve have a maximum, and how can this be explained? (cf. Bossel 1994: 429-432, Bossel 2007: 136-140).
7. What influence do different values of the damping parameter SELFCOUPLING FACTOR 2 $= d$ have on the shape of the frequency response?

References

Luenberger, D. G. 1979: *Introduction to Dynamic Systems – Theory, Models, and Applications*. John Wiley, New York.
Bossel, H. 1994: *Modeling and Simulation*. A K Peters, Wellesley MA (387-442).
Bossel, H. 2007: *Systems and Models - Complexity, Dynamics, Evolution, Sustainability*. Books on Demand, Norderstedt (124-153).

Z115 State space diagram

Simulation task

Quite generally, the development of the state of a dynamic system depends on its initial state. It may have been brought to this state by an exogenous input, i.e. an initial pulse. How the system subsequently develops – starting with the given initial conditions and without any further exogenous input – is of particular interest. It may drift toward a stable equilibrium state, or move away from the initial state in an unstable motion. Several very different, stable and unstable possibilities of dynamic behavior became evident in the investigations with model Z114 "Linear oscillator". The development of the system state as a function of the initial state becomes particularly obvious from the respective state trajectories, which – for two-dimensional systems – can be represented in a state diagram (phase plot). (For systems of higher dimension, the two-dimensional phase plot represents a projection of the system state on the plane defined by two particular states.)

Linear systems possess a single equilibrium point in state space. A stable system moves toward this point; an unstable system moves away from it. The behavior of linear systems is surprise-free, and can be explain by simple analysis. By contrast, nonlinear systems can have several stable and/or unstable equilibrium points, and furthermore exhibit still other perplexing phenomena. Usually their mathematics cannot be treated analytically, and comprehensive investigation of their behavior requires a large number of simulations (cf. Bossel 1994 and 2007, and simulation models in particular in Bossel 2007 *System Zoo 1*).

To get an overview of the state trajectories arising from all possible initial conditions, many simulations must therefore be presented in a comprehensive picture. (Unfortunately, the lucid representation of trajectories in state space is limited to two and three-dimensions.) Such an investigation would be too laborious if for every new simulation run initial conditions would have to be specified and programmed individually. It then makes sense to have the computer carry out the systematic variation of initial conditions, the corresponding simulations, and the complete presentation of the results. Standard simulation systems, however, usually do not provide this possibility.

An auxiliary module is documented in the following which can be coupled to a simulation model to produce state diagrams in two-dimensional state space with a grid of 10 by 10 = 100 initial states. This module is coupled to model Z114 "Linear oscillator" to produce state diagrams of its behavior for different combinations of coupling parameters. In Part 2 "Physics and engineering" and in other models of the System Zoo the module is used to clarify the behavior of particular (two-dimensional) nonlinear systems.

Simulation program

The complete model (oscillator + state diagram) is documented in Figure Z115a and the following model equations. Model Z114 "Linear oscillator" is used here as an example to demonstrate the application of the procedure. In the same way other models can be linked to the state diagram module. The coupling is effected by "shadow" or "ghost" variables shown in <pointed> brackets in the simulation diagram.

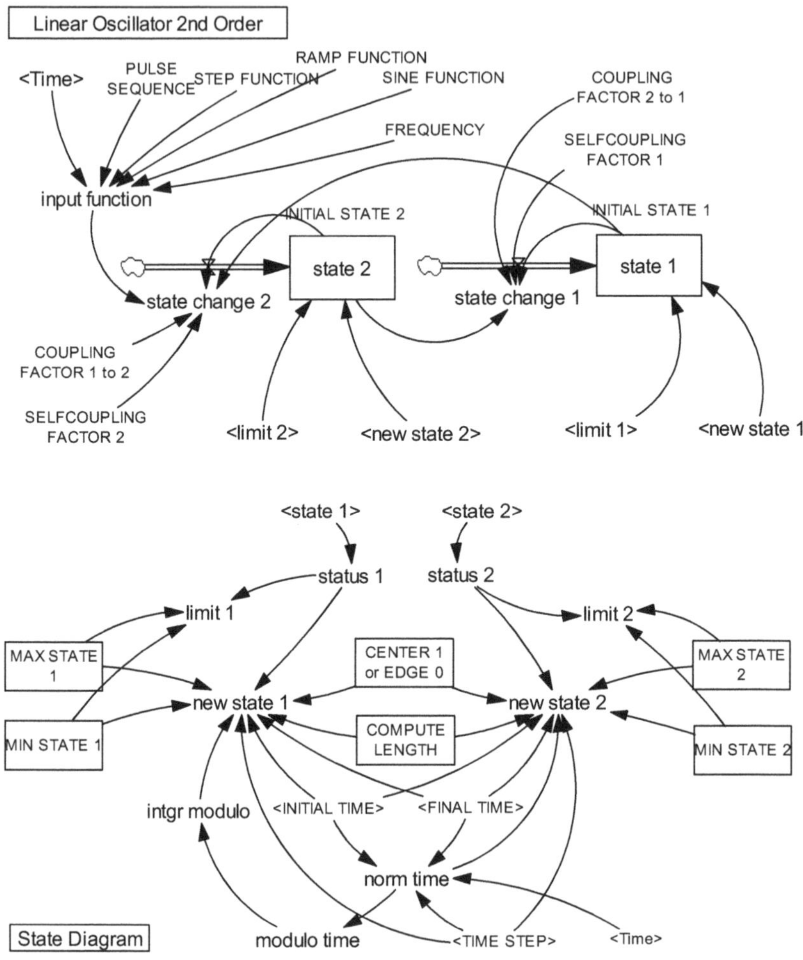

Figure Z115a: Simulation diagram for model Z114 "Linear oscillator" (upper part) connected to module Z115 "State space diagram" (lower part) for computation of global behavior. Both models are linked by "ghost" or "shadow" variables (e.g. <state 1>).

To compute consecutively the state trajectories of 10 by 10 = 100 initial states, *state 1* and *state 2* are reset to the new initial conditions before each new trajectory computation. This is repeated 100 times, and results are stored. Following these computations, the results are plotted in the state diagram. *Important note*: Choose pointwise plotting of results in standard simulation software ("dots" in the definition of "graph" in VenPLE/Vensim, "scatter plot" in Stella/iThink).

The system model is linked to the state diagram module by coupling model variables *state 1* and *state 2* to the module, and by coupling module quantities *limit 1, new state 1, limit 2,* and *new state 2* to the simulation model. The two equations for the computation of *state 1* and *state 2* in the simulation model must be augmented correspondingly (see **bold** terms in the model equations). The initial states are set to zero (and changed by the program during the run).

Before the simulation run, the six parameters shown in boxes (lower part of Figure Z115a) must be specified. Four of them define the range of initial states, i.e. the lower and upper limits of the range of initial conditions MIN STATE 1, MAX STATE 1, MIN STATE 2, MAX STATE 2. If the PARAMETER CENTER 1 OR EDGE 0 is chosen as "0", the initial states are placed at the intersections of the (10 by 10) grid lines, beginning at the lower left corner. For the choice "1", the initial states are placed at the center of each of the 10 by 10 fields. Finally, the time period COMPUTE LENGTH [*time unit*] for an individual trajectory simulation must be chosen. The total simulation period (FINAL TIME – INITIAL TIME) must be specified with a length of (at least) 100 times COMPUTE LENGTH. If the state trajectory extends significantly beyond the specified state region, the program terminates the computation of this trajectory and jumps to the next initial state.

SIMULATION MODEL (Oscillator)

Parameters and initial states
INITIAL STATE 1 = 0 [amount*Second]
INITIAL STATE 2 = 0 [amount]
SELFCOUPLING FACTOR 1 = 0 [1/Second]
SELFCOUPLING FACTOR 2 = -1 [1/Second]
COUPLING FACTOR 1 to 2 = -1 [1/(Second*Second)]
COUPLING FACTOR 2 to 1 = 1 [1]
PULSE SEQUENCE = 0 [amount/Second]
STEP FUNCTION = 0 [amount/Second]
RAMP FUNCTION = 0 [amount/Second]
SINE FUNCTION = 0 [amount/Second]
FREQUENCY = 0.1 [1/Second]

Dynamics
input function = PULSE SEQUENCE *50 *PULSE TRAIN (1, 0.02, 100, 100) +STEP FUNCTION *STEP (1, 1) +RAMP FUNCTION *RAMP (1, 1, 5) +SINE FUNCTION *SIN (2 *3.14159 *FREQUENCY *Time) [amount/Second]
state change 1 = SELFCOUPLING FACTOR 1 *state 1 +COUPLING FACTOR 2 to 1 *state 2 [amount]
state change 2 = COUPLING FACTOR 1 to 2 *state 1 +SELFCOUPLING FACTOR 2 *state 2 +input function [amount /Second]
state 1 = INTEG ((+state change 1) **limit 1 +new state 1**, INITIAL STATE 1) [Second*amount]
state 2 = INTEG ((+state change 2) **limit 2 +new state 2**, INITIAL STATE 2) [amount]
   {*model state variables state 1 and state 2 are coupled to the state diagram calculation. limit 1, new state 1, limit 2, new state 2 are computed in the state diagram calculation and coupled to the simulation model.*}

Simulation time parameters
INITIAL TIME = 0 [Second]
FINAL TIME = 100 [Second] {*use 100 times trajectory computation period!*}
TIME STEP = 0.05 [Second]
COMPUTE LENGTH = 1 [Second] {*computation period for individual trajectory*}

STATE DIAGRAM

Domain of investigation
MAX STATE 1 = 1
MIN STATE 1 = -1
MAX STATE 2 = 1
MIN STATE 2 = -1
center 1 or EDGE 0 = 1 {*state diagram in center: 1, at left lower corner: 0*}

State trajectories
status 1 = **state 1** [amount*Second] {*linkages of state variables*}
status 2 = **state 2** [amount]
norm time = (Time +TIME STEP/2) /(FINAL TIME -INITIAL TIME)
modulo time = MODULO (10*norm time, 1)
intgr modulo = INTEGER (10 *modulo time)
limit 1 = IF THEN ELSE (status 1 > (3 *MAX STATE 1 -MIN STATE 1) /2, 0, IF THEN
      ELSE (status 1 < (3*MIN STATE 1 -MAX STATE 1) /2, 0, 1))
limit 2 = IF THEN ELSE (status 2 > (3 *MAX STATE 2 -MIN STATE 2) /2, 0, IF THEN
      ELSE (status 2 < (3*MIN STATE 2 -MAX STATE 2) /2, 0, 1))
new state 1 = PULSE TRAIN (INITIAL TIME, TIME STEP, COMPUTE LENGTH, FI-
      NAL TIME) *((MAX STATE 1 -MIN STATE 1) *(intgr modulo/10) +MIN STATE 1
      +center 1 or EDGE 0 *(MAX STATE 1 -MIN STATE 1) /20 -Status 1) /TIME
      STEP
new state 2 = PULSE TRAIN (INITIAL TIME, TIME STEP, COMPUTE LENGTH, FI-
      NAL TIME) *(((MAX STATE 2 -MIN STATE 2) /10) *INTEGER (10*norm time)
      +MIN STATE 2 +center 1 or EDGE 0 *(MAX STATE 2 -MIN STATE 2) /20 -
      Status 2) /TIME STEP

{*IMPORTANT: Plot results as **dots**! Select proper diagram definition:*
in VenPLE/Vensim select "dots", in Stella/iThink select "scatter plot"}

Simulation results

The state diagrams calculated with the present program for model Z114 "Linear oscil-
lator" with initial values in the range $-1 < z_1 < 1$ und $-1 < z_2 < 1$ are shown in Figures
Z115b-e. In these cases only the COUPLING FACTOR 1 to 2 (= c) and the SELFCOU-
PLING FACTOR 2 (= d) were changed; other values are those of the default setting.

    Figure Z115b shows a stable *focus* ($c = -1$, $d = -1$), Figure Z115c a stable *node*
($c = -1$, $d = -2$), Figure Z115d a *center* (vortex) ($c = -1$, $d = 0$), and Figure Z115e an
(unstable) *saddle* ($c = 1$, $d = -1$). In each case the equilibrium point is at the coordinate origin (0, 0). For stable motion, the trajectories move towards this point, for unstable motion they move away from it. (The direction of the motion cannot be inferred
from these state diagrams, but it follows from the time diagrams.)

Exercises

1. Produce state diagrams for other system parameters (change primarily c and d).
2. How do the system parameters have to be chosen to cause an unstable focus and an
unstable node which have the same state diagrams as the (stable) focus and node in
Figures Z115b and c (but opposite direction of motion)?

3. Link one or several of the two-dimensional nonlinear system models from Part 2 "Physics and engineering" to the state diagram module, and examine and discuss the results, particularly the location of the equilibrium points and their (in)stability, the direction of motion of the state trajectories, and other phenomena like limit cycles or chaos.

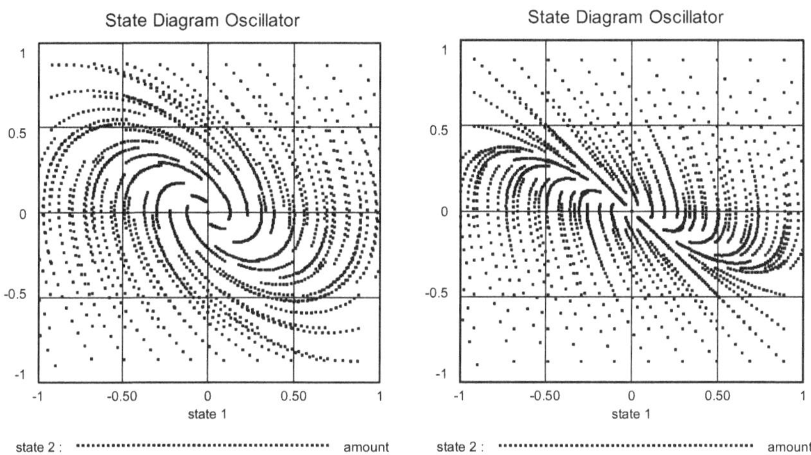

Figure Z115b: Stable focus ($c = -1$, $d = -1$).
Figure Z115c: Stable node ($c = -1$, $d = -2$).

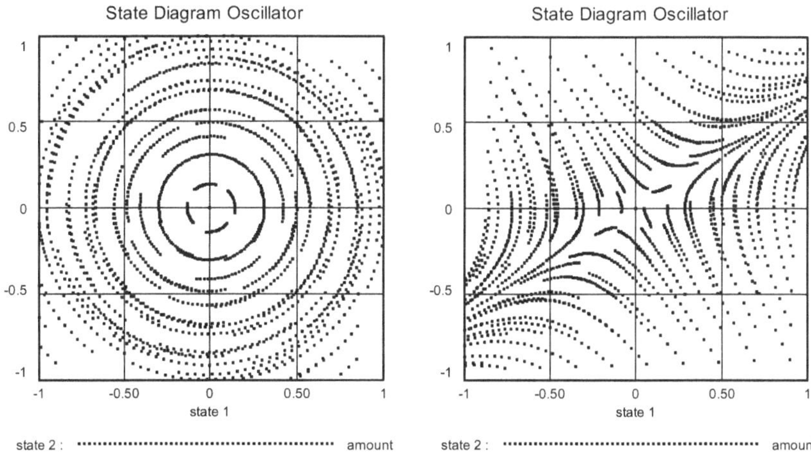

Figure Z115d: Center (vortex) ($c = -1$, $d = 0$).
Figure Z115e: Saddle (always unstable) ($c = 1$, $d = -1$).

Z116 Triple integration and exponential delay

Simulation task

The delay effect of exponential feedback was already treated in models Z104 "Exponential delay" and Z112 "Double integration and exponential delay". Every additional integrator with negative self-coupling delays the input signal further. Exponential delays of first and third order (with three integrators) are frequently used in simulation models and are an integral part of standard simulation software (as DELAY1 and DELAY3).

For such delays, the same self-coupling factor is used for each integrator. Since it has a negative sign and therefore causes a constant decrease of state level ("leakage") the self-coupling factor is designated here as SPECIFIC LOSS RATE a. If $a = 0$, the system produces a triple integration of the input signal. The result then corresponds to analytical integration formulae. If the loss rate $|a| > 0$, then a delay effect of value $1/a$ arises at every integrator. Altogether, the delay time is therefore $T_{D3} = 3/a$ [*time unit*]. For strong feedback a short signal delay arises, for weak feedback the delay effect is larger.

Simulation model

The model is documented in the simulation diagram of Figure Z116a and in the following equations. The first integrator integrates the *change Z*, i.e. *input function* reduced by *state Z* * SPECIF LOSS RATE. The same process occurs at the other integrators. The SPECIF LOSS RATE a is identical for all three integrators. A combination of test functions is again used as *input function* (cf. previous models).

Parameters (*initial states = 0, cf. INTEG equations*)
SPECIF LOSS RATE = 1 [1/Second]
PULSE SEQUENCE = 0 [1/Second]
STEP FUNCTION = 1 [1/Second]
RAMP FUNCTION = 0 [1/Second]
SINE FUNCTION = 0 [1/Second]
FREQUENCY = 0.1 [1/Second]

Dynamics
change X = state Y -SPECIF LOSS RATE *state X [Second]
change Y = state Z -SPECIF LOSS RATE *state Y [1]
change Z = input function -SPECIF LOSS RATE *state Z [1/Second]
input function = PULSE SEQUENCE *50 *PULSE TRAIN (1, 0.02, 50, 50) +STEP
        FUNCTION*STEP(1, 1) +RAMP FUNCTION *RAMP (1, 1, 5) +SINE FUNC-
        TION *SIN(2 *3.14159 *FREQUENCY *Time) [1/Second]
state X = INTEG (change X, 0) [Second*Second]
state Y = INTEG (change Y, 0) [Second]
state Z = INTEG (change Z, 0) [1]

Simulation time parameters
INITIAL TIME = 0 [Second]
FINAL TIME = 10 [Second]
TIME STEP = 0.01 [Second]

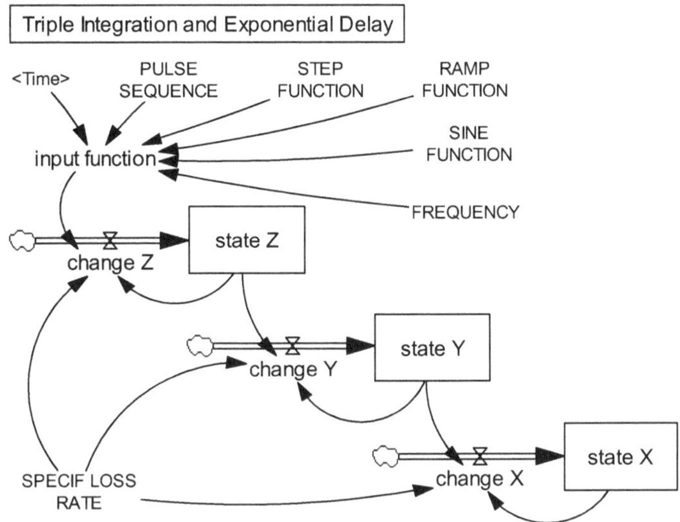

Figure Z116a: Simulation diagram for triple integration with damping.

Simulation results

Figure Z116b shows the result for the normal triple integration process ($a = 0$). In this case the input function is a single pulse with pulse height 50, defined by PULSE TRAIN (1, 0.02, 50, 50), i.e. the pulse beginning at *time* = 1, pulse duration = 0.02, pulse period = 50, end of the pulse sequence = 50. The *pulse strength (pulse area = pulse height of * pulse width)* is here = 1. This leads to a constant output signal of 1 for *state Z*. In the second integrator, this becomes a linear increase of *state Y* = 1 · *time* . In the third integrator this is integrated to a quadratic function of time *state X* = 1 · *time*^{2} . The simulation result corresponds to the integration formulae.

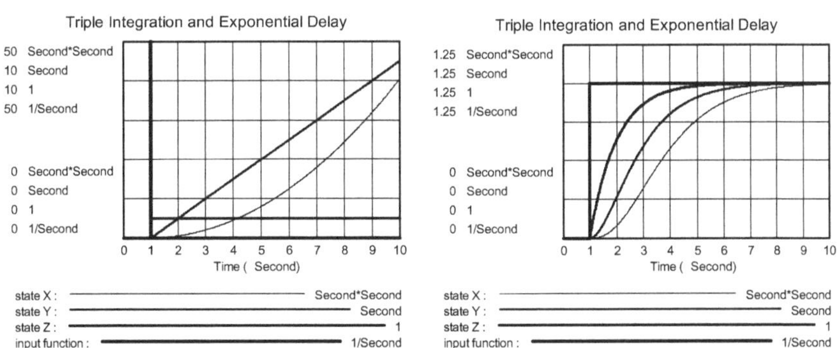

Figure Z116b: Triple integration of a pulse (without damping) leads to step, ramp, and parabola after the consecutive integrations.
Figure Z116c: A delay effect arises for negative feedback (damping, leak).

Figure Z116c shows the result for negative feedback (positive SPECIFIC LOSS RATE $a > 0$, here $a = 1$). At each integrator a state-proportional loss arises now. If $u(t)$ is a step function with $u = \text{const}$ (here $u = 1$) for $t > t_{\text{step}}$ (here $t_{\text{step}} = 1$), the *state Z* increases up to the point where the exponential loss rate corresponds exactly to the input signal u. As input to the integrator for *state Y*, the now constant output signal (*state Z*) undergoes the same effect, eventually also producing a constant output *state Y*. The process repeats again in the integrator for *state X*. For constant input u the equilibrium values at the three integrators become: $z^* = u/a$, $y^* = u/a^2$, $x^* = u/a^3$.

The delay effect becomes even more obvious in Figure Z116d. Here a step of strength = 1 at *time* = 1 is superimposed on a sinusoidal oscillation starting at *time* = 0 with FREQUENCY $f = 0.5$. This step leads at the first integrator to an increase of the *state Z* level by 1, which now also moves the mean output value of the other integrators to this value. At all three integrators a permanent oscillation around the state value = 1 develops.

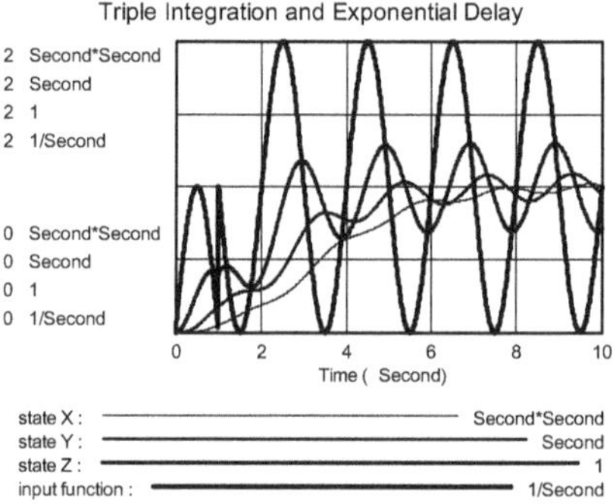

Figure Z116d: Delay of a sinusoidal oscillation with a superimposed step (at *time* = 1).

Exercises

1. Use the model with $a = 0$ to integrate the different test functions (pulse, step, ramp, sine) over time once, twice, and three times. Compare the results with the analytical integration formulae.

2. Investigate the dependence of the equilibrium values of the three integrators on feedback parameter a.

3. Compute analytically by what factor the amplitude of a sinusoidal oscillation is changed with each integration. Check the result by simulation.

4. Confirm by simulations that (for linear systems like the present one) the sum of output states in response to individually acting input functions is equal to the output state resulting from an input function which is the sum of the individual functions (*superposition principle*, cf. Bossel 2007: 137). [I.c. (a) adjust the input parameters to construct a complex input function containing several of the functions provided, and compute the output (*state X*). Then (b) run individual simulations for each of the functions used (using the same parameters), note the individual outputs (*state X*), then add them up and compare the result with (a)].

Z117 Linear oscillator of third order

Simulation task

Linear systems of order greater than two capable of oscillation are found in many technical systems, in particular in control systems. Do we have to expect qualitatively different behaviors of such systems? This is not the case: All behavioral possibilities of linear systems already appear in linear systems of second order. Their investigation therefore provides a fundamental understanding of the behavior of linear systems of *any* order (with more than two state variables). (For the theoretical background, see e.g. Bossel 2007: 124-140, Bossel 1994: 388-432.)

Linear systems producing identical behavior can be realized in different ways. Their differential equation has the (vector) form

$$\frac{d\mathbf{x}}{dt} = \mathbf{A}\mathbf{x}$$

where $\mathbf{x}$ is the state vector and $\mathbf{A}$ the system matrix. The system matrix reflects the couplings between the state variables: the more non-zero entries it contains, the more links exist between state variables. One distinguishes three different types of system matrices: the *general form*, the *standard form* and the *normal form*. Each of these forms can be mathematically converted into every other one.

The *general form* can have an arbitrary distribution of matrix coefficients (indicating state variable connections). For N state quantities N^2 connections are therefore possible. In the *normal form* of the system matrix, the behavioral modes of the system are separated; each state variable is therefore coupled only to itself. The number of possible connections is therefore reduced to N. This type of system structure is mathematically elegant, but difficult to achieve in technical solutions in practice. The *standard form*, however, has particular practical relevance because it corresponds to the normal coupling of technical components. It represents $(2 \cdot N - 1)$ possible connections and the structural complexity is therefore reduced to an acceptable level (cf. Bossel 1994: 412-416).

For a system of third order the system matrix has the *standard form*

$$\mathbf{A} = \begin{bmatrix} 0 & 1 & 0 \\ 0 & 0 & 1 \\ a & b & c \end{bmatrix}$$

This system matrix has the characteristic polynomial

$$-\lambda^3 + c\lambda^2 + b\lambda + a = 0$$

Its solution yields the eigenvalues λ_1, λ_2, λ_3 of the system. In general, these are complex numbers with real and imaginary parts, i.e. $\lambda_i = \mathrm{Re}(\lambda_i) + \mathrm{Im}(\lambda_i)$, or $\lambda_i = \sigma_i \pm i\omega_i$. The behavioral possibilities of this system are identical to those of the oscillator of second order (damped and undamped, periodical and aperiodic behavior). The parameter c controls primarily the damping, the parameters a and b are responsible for the oscillation and its frequency. All other linear systems of third order can also be represented in this form. In the standard form all state variables are coupled back to the first state variable (z) of the integration chain. In addition, an exogenous function $u(t)$ acts on the state quantity z.

Three eigenvalues and three behavior modes of the type $e^{\lambda t}$ or $e^{\sigma t} \sin \omega t$ correspond to the three state variables. Here σ is the real part of the (generally complex) eigenvalue: $\sigma = Re(\lambda)$. The general solution is a combination of these periodical, aperiodic, damped or amplified solutions. The system behavior is stable only if the real part of the eigenvalue is negative, i.e. $Re(\lambda) < 0$. The location of the eigenvalues in the complex plane determines the behavior of the system in the same way as for the harmonic oscillator of second order.

Simulation model

The model for the oscillator of third order with coupling coefficients corresponding to the standard form of the system matrix is documented in Figure Z117a and the following model equations. As evident from the system matrix, the *state Z* is coupled to itself and the other state variables, while *state Y* and *state X* are merely integrating the preceding state variable. The examination of possible behaviors therefore reduces to variation of the three COUPLING FACTORS Z to Z ($= c$), Y to Z ($= b$) and X to Z ($= a$). The usual test functions are provided; they can be combined to generate a large variety of *input functions*.

Parameters and initial values
INITIAL VALUE X = 0 [Second*Second]
INITIAL VALUE Y = 0 [Second]
INITIAL VALUE Z = 0 [1]
COUPLING FACTOR X TO Z = -1 [1/(Second*Second*Second)]
COUPLING FACTOR Y TO Z = -1 [1/(Second*Second)]
COUPLING FACTOR Z TO Z = -2 [1/Second]
PULSE SEQUENCE = 1 [1/Second]
STEP FUNCTION = 0 [1/Second]
RAMP FUNCTION = 0 [1/Second]
SINE FUNCTION = 0 [1/Second]
FREQUENCY = 0.1 [1/Second]

Dynamics
input function = PULSE SEQUENCE *10 *PULSE TRAIN (1, 0.1, FINAL TIME, FINAL
    TIME) +STEP FUNCTION *STEP(1, 1) +RAMP FUNCTION *RAMP (1, 1, 5)
    +SINE FUNCTION *SIN (2 *3.14159 *FREQUENCY *Time) [1/Second]
change X = state Y [Second]
change Y = state Z [1]
change Z = (COUPLING FACTOR X TO Z *state X) +(COUPLING FACTOR Y TO Z
    *state Y) +(COUPLING FACTOR Z TO Z *state Z) +input function [1/Second]
state X = INTEG (change X, INITIAL VALUE X) [Second*Second]
state Y = INTEG (change Y, INITIAL VALUE Y) [Second]
state Z = INTEG (change Z, INITIAL VALUE Z) [1]

Simulation time parameters
INITIAL TIME = 0 [Second]
FINAL TIME = 20 [Second]
TIME STEP = 0.01 [Second]

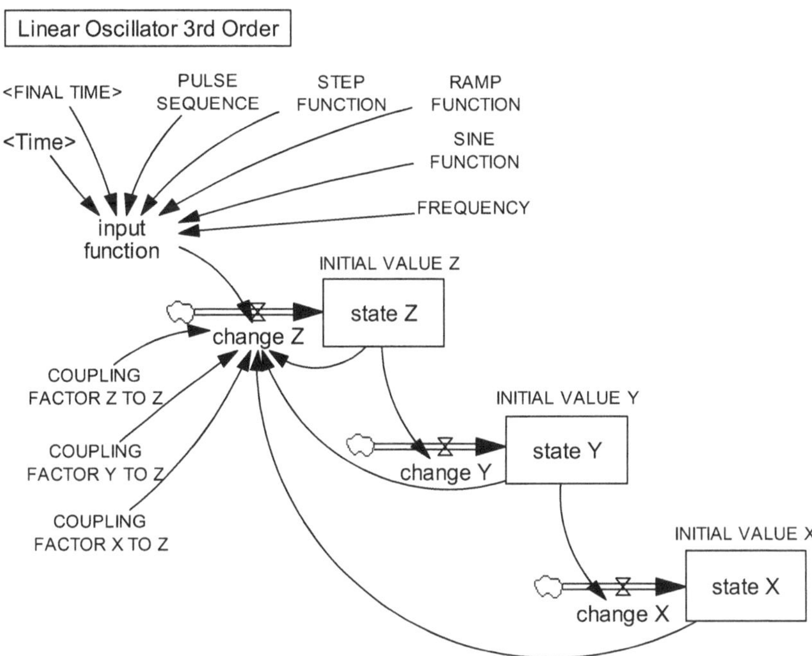

Figure Z117a: Simulation diagram of the linear oscillator of third order.

For the default parameter settings ($a = -1$, $b = -1$, $c = -2$) a strongly damped oscillation arises after applying a pulse of magnitude 1 to the system at rest at time $t = 1$ (Figure Z117b). A period of about 8 [time units] can be read from the time diagram. In each integration step, the respective input signal is transformed to a delayed output signal.

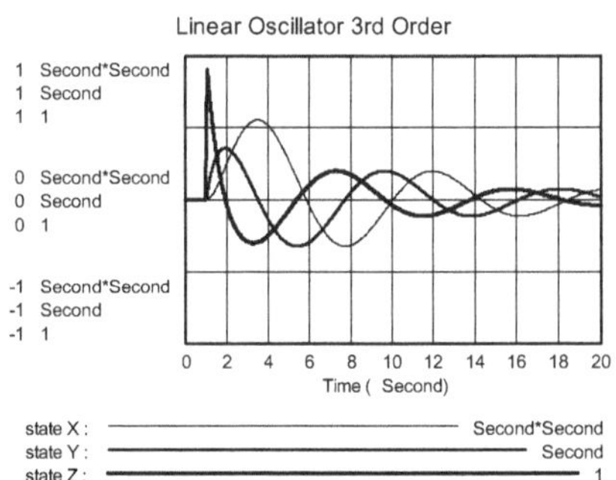

Figure Z117b: Oscillations of the three state variables as a result of an initial pulse.

Exercises

1. Run multiple simulations with different parameter values, plot the resulting time diagrams of a state variable in a common diagram, and search for parameter constellations leading to strongly damped solutions. Find a parameter combination which causes the oscillations to disappear.
2. How can the frequency and the period of oscillation be changed by changing parameters? Which parameters have what effect?
3. Using the parameters of the default setting, calculate the eigenvalues, damping and natural frequency of the system. Compare with the simulation result.
4. Find a parameter combination for an aperiodically damped solution (corresponding to three purely real and negative eigenvalues).
5. Find combinations of coupling parameters which cause each of the qualitatively different, stable and unstable behaviors generated with model Z114 (and Z115).

References

Textbooks on control system theory, also:
Bossel, H. 1994: *Modeling and Simulation*. A K Peters, Wellesley MA (387-442).
Bossel, H. 2004: *Systems and Models – Complexity, Dynamics, Evolution, Sustainability*. Books on Demand, Norderstedt (124-153)

2
Physics and Engineering

Overview

Independently of each other Newton and Leibniz developed the infinitesimal calculus at almost the same time, thus laying the foundation for the computation of dynamic systems. Basic processes of mechanics, like the movement of bodies in response to accelerations, could be described and calculated exactly using differential and integral calculus. Most mathematical models and simulations have been developed for dynamic systems in engineering and physics because their dynamics is largely determined by laws of nature which can be formulated as exact mathematical expressions. The methods of system representation, systems analysis, control and optimization, computation and computer simulation developed in engineering and physics in the latter 20th century have increasingly been transferred into other, i.e. "soft" areas of the natural and social sciences.

Before the advent of computers, investigations of dynamic systems were limited almost exclusively to "linear" or "linearized" dynamic systems, where the computation effort could be handled with pen, paper and much time. Such systems, however, have only a limited behavioral repertoire. The enormous variety of dynamic behaviors became evident only after electronic computers made routine simulations of "nonlinear" systems possible – which previously could not be studied because of the extensive manual calculations required. In particular, it became evident that nonlinear systems may show chaotic behavior (as in weather dynamics) on the one hand, or self-organizing and self-ordering behavior (as in laser dynamics) on the other hand. Nonlinear processes lead – among significant other effects – to the development of life and its evolution, as well as similar processes in the social sciences. So it is nonlinear systems which mainly determine developments in all domains of reality.

The 14 simulation models introduced in this part of the System Zoo provide on the one hand insights into possible behaviors of nonlinear systems. On the other hand they also represent typical technical applications of simulation models in tasks of control and optimization and in the computation of heat flows and velocity distributions in fluid flows.

Z201 Rotating pendulum. A simpler mechanical system than a pendulum is hardly conceivable: A mass connected to a pivot by a rigid rod. Formulated correctly, it turns out to be a nonlinear system with different modes of behavior depending on initial speed and position of the pendulum and its air resistance. Its movement is unstable at the upper dead center, stable around the lower dead center. If the initial velocity is high, the pendulum will at first revolve around its pivot. As friction slows it down, it will eventually no longer reach the upper dead center, will swing back and forth around the lower dead center, and finally come to rest there. Only if the pendulum swings remain quite small can the process be formulated as a linear differential equation – which can also be used to obtain an approximate description of the behavior of the nonlinear system in the neighborhood of its dead centers.

Z202 Oscillator with limit cycle (van der Pol). In the early days of radio technology, the stabilization of radio frequencies was attained by a nonlinear component (a triode radio tube) in electronic circuits, which generated a constant frequency "limit cycle". The same system structure also stabilizes cardiac activity and causes wind-excited vibrations of large structures, wing flutter of airplanes or vibrations in certain chemical reactions.

Z203 Brusselator. In chemistry the Belousov-Zhabotinsky reaction has become well known for its ability to cause oscillations and spatial patterns under certain conditions. This is caused by a nonlinear change of state. The "Brusselator" is a simple model of this reaction, which allows studying the different possible behaviors. The chemical oscillation, i.e. the periodical fluctuation of the concentrations of two components, is in this case also stabilized by a limit cycle.

Z204 Bistable oscillator. If a cubic negative feedback of one state variable to another is added to a linear oscillator (like Z114), this may lead to (damped) oscillations around one of two stable equilibrium points. A (positive or negative) cubic feedback term is characteristic of the so-called Duffing systems. Examples of such systems are the electrical relaxation circuit with a cubic resistance function, a mechanical oscillator with progressive spring stiffness, and a leaf spring between two permanent magnets. The oscillator with a negative cubic feedback can be used as a flip-flop switch for sorting signals: Depending on initial state the system approaches one of two possible stable states.

Z205 Chaotic bistable oscillator. The regular behavior of the bistable oscillator changes completely if it is excited by a harmonic oscillation. The system can be realized, for example, by a solid frame in which a leaf spring can swing between two permanent magnets. If the frame is moved back and forth at a given frequency, the leaf spring will oscillate for a while around one of the equilibrium points of the bistable oscillator before jumping "completely unpredictably" to the region of attraction of the other equilibrium point, remaining there for a while, jumping back to the other region of attraction etc. The system exhibits unforeseeable chaotic behavior although its components (the bistable oscillator and the periodic excitation) are completely regular and exactly calculable processes.

Z206 Heat, weather, and chaos (Lorenz system). Chaotic events stand out by the fact that tiny differences of initial states can lead to completely different system developments. The weather is an example: The proverbial flutter of a butterfly in West Africa can (theoretically) determine whether a hurricane develops later in the Caribbean. The meteorologist Lorenz encountered this phenomenon in his computer simulations of fundamental weather events with an extremely simplified system of equations. On the Lorenz attractor – which itself has the shape of a butterfly – the system state moves on an unpredictable trajectory around one, then around the other equilibrium point, then jumps back to the first, without ever coming to rest.

Z207 Chaotic attractor (Rössler). Fundamental insights into chaotic processes can be obtained from a study of the Rössler attractor, the "simply folded ribbon". This is a mathematical construction and does not represent any real system. Depending on parameter choice, multi-period limit cycles or similar (but nonrepeating) chaotic trajectories develop.

Z208 Coupled dynamos and chaos. If two dynamos are coupled so that the current produced in one dynamo drives the other by an electric motor and *vice versa*, and friction losses are neglected, this coupled system also produces chaotic behavior. The state trajectories move around one equilibrium point before suddenly jumping into the region of attraction of the other equilibrium point.

Z209 Balancing an inverted pendulum. Unstable systems can be stabilized by the influence of additional system components. For example, a broom can be stabilized upright on a finger by moving the finger back and forth in response to the motion of the broom to keep it from tipping. The problem has great practical importance e.g. for launching of space vehicles (or controlling the Segway vehicle). To control a process effectively, important state variables must be measured and compared with the desired state. If a deviation occurs, a negative feedback produces a force counteracting the unwanted development. The design of efficient control systems is a science by itself because each one must be developed specifically for a given system and its particular application. In this design process computer simulation plays an important role. Negative feedbacks for the control of system states are not only used in technical processes but they are also found almost everywhere else, especially also in organisms (e.g. temperature control) or in ecosystems (e.g. population control by nutrient deficiencies or predators).

Z210 Optimizing glider flight: search for thermals. The stabilization of unstable processes is only one of the tasks for which control systems are developed. In particular, optimization of processes is an important application: deriving maximum utility (room heat, transportation, production) with the least possible energy consumption; maximum harvest for the least possible amounts of irrigation, fertilizers and pesticides; or in general: best possible use of available resources. Because of the complexity of the system and the circumstances under which it shall optimally operate, finding possible optima is often the most difficult part of the task. Optimization compares to the task of not only finding quickly– while blindfolded – the highest mountain in mountainous terrain, but also to climb it as rapidly as possible. This task also faces the glider pilot who must find invisible updrafts along his route and climb in them as quickly as possible. An optimal flight rule has to be found to locate updrafts ("thermals") quickly and use them efficiently. Interestingly enough, the flight rule found is also applicable to completely different cases where a search under unknown conditions must be optimized.

Z211 Flight dynamics. Pilots of civil and military aircraft complete hundreds of flight hours in flight simulators. Their simulated dynamics correspond exactly to that of real aircraft so that the flying experience gained in the simulator can replace many costly training flights on actual aircraft. In particular, it is possible to practice dangerous and risky maneuvers which cannot be safely practiced in actual flights. At the heart of every flight simulator is a system of differential equations which allows computing the movements of the airplane in response to the pilot's control inputs and the resulting velocities and accelerations, and the forces and moments caused by them. Mass forces and mass moments as well as aerodynamic forces and moments of fuselage, wings and control surfaces must be accounted for in this system of equations. The general system of flight dynamic equations must therefore be quantified by a large number of aircraft-specific coefficients which in turn stem from extensive calculations or measurements. The simulation of flight dynamics is an important part of aircraft design to obtain well-balanced handling characteristics and high maneuverability with a high margin of flight safety.

Z212 House heating dynamics. The temperature inside a residential building is dependent on how much heat is produced by the heating system, gained by sun irradiation through windows, and lost to the environment through the roof, windows, outer walls, and floors. The area of external surfaces, their orientation to the sun and the heat transfer values of construction and insulation materials used play an outstanding role in the overall heat balance. Also, the heat balance depends on the daily change of outside temperature, heater performance, position of the sun and sunshine duration. To design and equip a house optimally for low energy consumption, its seasonally changing energy balance must be represented in a simulation model. With such a model very different design concepts can easily be investigated to find an optimal layout with respect to window areas, building materials, heating system, energy consumption, costs and esthetics.

Z213 Integral relations and heat conduction. The systems considered so far have "time" as the only independent variable. They can therefore be expressed by ordinary differential equations in which only derivatives with respect to time appear (e.g. df/dt). However, many processes in the real world are not only time-variable (such as outside temperature of a house, speed of a vehicle, strength of an electric current) but they also have a spatially variable distribution (for example, temperature distribution in a heated piece of steel, aerodynamic flow field around a wing, concentration in a river of a chemical introduced from a point source). Such processes must be described by partial differential equations, in which partial derivatives (symbolized by ∂) with respect to all of the independent variables (mostly time t and spatial coordinates x, y, and z) appear (i.e. $\partial f/\partial t$, $\partial f/\partial x$, $\partial f/\partial y$, $\partial f/\partial z$ etc.). With some mathematical effort, it is sometimes possible to replace a process represented by partial differential equations with a system of ordinary differential equations which can then be calculated efficiently with the computer procedures developed for these. The trick is to approximate the spatial distribution by a mathematical function whose parameters can then be found by numerical integration of the remaining ordinary differential equations. The "method of finite elements" widely used in technical applications is a variant of this procedure (e.g. stress and temperature distributions in complex structures, flow calculations, simulation of crash tests). As a simple example of this procedure the temperature distribution in a metal bar with time-variant heating (or cooling) at each end is computed as function of time and location.

Z214 Boundary layer flow. Mathematically the spatial dependence ($\partial f/\partial x$) can be treated in the same way as the temporal dependence ($\partial f/\partial t$). With the integral procedure of model Z213 it is therefore possible to calculate stationary (time-independent) processes originally expressed by partial differential equations in two space dimensions. Two-dimensional flows, for example around aircraft wings are an example. With such computations airfoils can be developed which have good lift values at minimal drag. The procedure is demonstrated by computing the boundary layer flow around a circular cylinder. In this case the flow separates behind the point of maximum thickness, forming a turbulent wake.

Z201 Rotating pendulum

Simulation task

A pendulum is a rather simple physical system. Its equations of motion are nonlinear, however, and it therefore exhibits rather complex behavior. It can swing back and forth (thereby constantly changing its direction of motion) or circle around its mounting point (maintaining its direction of motion). It has equilibrium points which can be stable (at the lower dead center) or unstable (at the upper dead center). Its motion depends decisively on initial conditions (initial angle and angular velocity).

The behavior can be correctly described by an idealized model system. Assumptions: A weightless, rigid rod supporting the pendulum mass at one end is mounted at its other end on a horizontal axis allowing it to rotate in a vertical plane. If the pendulum starts initially with high angular velocity, it can repeatedly rotate around the pivot before it comes to rest at the lower dead center after swinging back and forth several times in damped motion. The motion is damped by air friction of the pendulum; gravity causes it to move towards the lower dead center.

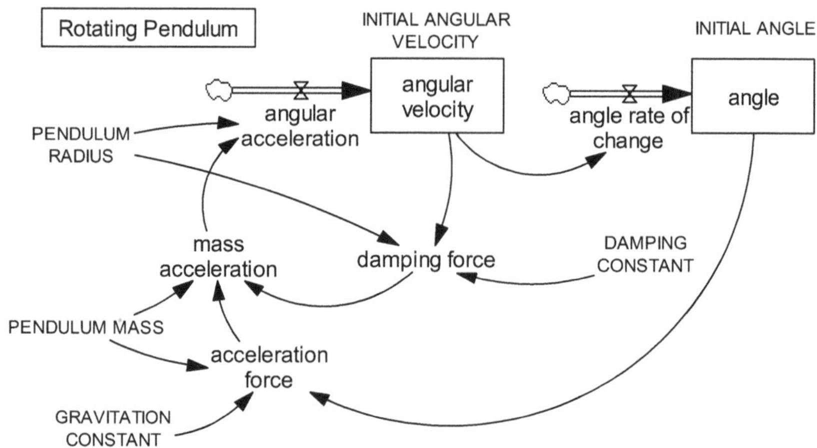

Figure Z201a: Simulation diagram for the rotating pendulum.

Simulation model

The model of the rotating pendulum is developed in detail elsewhere (Bossel 1994: 126-135, 172-195; Bossel 2007: 108-111, 119-122). Figure Z201a shows the simulation diagram; the corresponding model equations are listed in the following. The system with state variables *angular velocity* and *angle* has a feedback loop connecting the two state variables, which means that oscillatory behavior is to be expected. The *angular velocity* is affected by a *damping force* proportional to DAMPING CONSTANT and to current *angular velocity* (laminar drag force). The *damping force* determines how quickly the motion comes to rest. Corresponding to the *angle* of the pendulum motion the acceleration component of gravitation in the direction of the trajectory (*accelera-*

tion force) changes nonlinearly with the sine of the *angle*. Since gravity always pulls "down", it accelerates the motion as the pendulum swings downward, and decelerates it as the pendulum swings upward. Besides the damping, the initial conditions, particularly the angular velocity, are of great importance for the subsequent motion.

Parameters and initial states
INITIAL ANGLE = 0 [1]
INITIAL ANGULAR VELOCITY = 10 [1/Second]
PENDULUM MASS = 1 [kg]
PENDULUM RADIUS = 1 [m]
DAMPING CONSTANT = 1 [kg/Second]   [N/(m/s) = (kg*m/s^2) /(m/s) = kg/s]
GRAVITATION CONSTANT = 9.81 [m/(Second*Second)]

Dynamics
acceleration force = PENDULUM MASS *GRAVITATION CONSTANT *SIN(angle)
      [m*kg/(Second*Second)]
damping force = DAMPING CONSTANT *angular velocity *PENDULUM RADIUS
      [m*kg/(Second*Second)]
mass acceleration = -(acceleration force +damping force) /PENDULUM MASS
      [m/(Second*Second)]
angular acceleration = mass acceleration /PENDULUM RADIUS [(1/Second) /Second]
angular velocity = INTEG (+angular acceleration, INITIAL ANGULAR VELOCITY)
      [1/Second]
angle rate of change = angular velocity [1/Second]
angle = INTEG (+angle rate of change, INITIAL ANGLE) [1]

Simulation time parameters
INITIAL TIME = 0 [Second]
FINAL TIME = 10 [Second]
TIME STEP = 0.01 [Second]

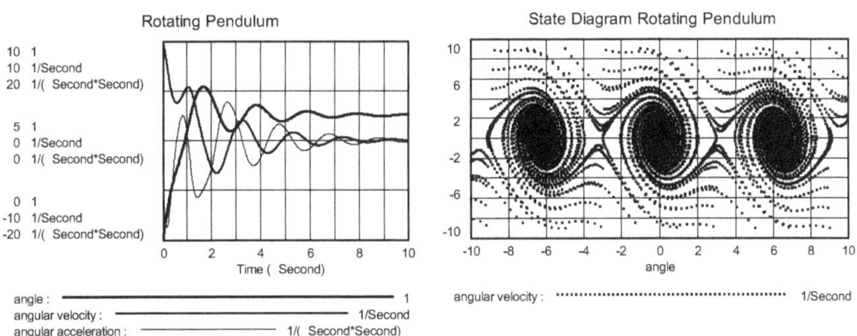

Figure Z201b: Time diagram for initially high angular velocity.
Figure Z201c: State trajectories around the upper dead center (unstable saddle) and lower dead center (stable focus).

Simulation results

Figure Z201b shows simulation results for the initial values and parameters chosen above. With this parameter setting the initial angular velocity is high enough to cause at first a full rotation of the pendulum, and then a strongly damped swinging motion. The pendulum comes to rest after one full rotation at an angle of $2\pi = 6.28 = 360$ degrees. The influence of different initial values on the motion, i.e. the global behavior for the chosen parameters can be examined by coupling the model to the auxiliary module Z115 "State space diagram" (Figure Z201c). For high initial *angular velocity*, the state trajectories show initial rotation (increasing *angle*), which slows down near the upper dead centers (unstable saddles). As the *angular velocity* decreases to zero, the pendulum reverses direction and begins to swing (oscillate) around the lower dead centers, where it finally comes to rest (stable focus). The pendulum dynamics is examined in more detail elsewhere (Bossel 1994: 176-195).

Exercises

1. Simulate the pendulum system with DAMPING CONSTANT $= 0.1$ and try to understand the state diagram (*angular velocity* over *angle*) in connection with the corresponding time diagrams for the different initial conditions. Identify the trajectories representing (a) swinging motion, and (b) rotation around the pivot point.
2. Change the DAMPING CONSTANT d in the range from 0 to 1 (negative values do not make physical sense). Interpret the state trajectories. What do you observe for $d = 0$ if you change the step width (TIME STEP) of the integration procedure (Euler or Runge-Kutta)? Which procedure and which step width should be employed to calculate the undamped rotation correctly?
3. Examine the map of state trajectories for different values of the DAMPING CONSTANT using the auxiliary module Z115 "State space diagram".
4. How does the pendulum frequency change as a function of PENDULUM RADIUS and PENDULUM MASS?
5. Examine and compare the time response for different DAMPING CONSTANT at initial angular velocities permitting at least one rotation even for the strongest damping value used. How can different equilibrium states for different damping be explained? (For this comparison, store the different runs under separate labels and then draw them in a common diagram.)
6. Linearize the state equations for the pendulum at each of the two equilibrium points (upper and lower dead center; cf. Bossel 1994: 190-195; Bossel 2007: 144-147). Determine the (coefficients of the) system matrix of the corresponding linearized system. Insert the coefficients found into the linear system of model Z114 "Linear oscillator", and compare the map of state trajectories for the nonlinear case (model Z201 coupled to module Z115) with the corresponding map for the linearized system (model Z114 coupled to module Z115) in the neighborhood of the saddle (upper dead center) and focus (lower dead center).

References

Bossel, H. 1994: *Modeling and Simulation*. A K Peters, Wellesley MA. (126-135, 141-146, 172-195)

Bossel, H. 2004: *Systeme, Dynamik, Simulation – Modellbildung, Analyse und Simulation komplexer Systeme*. Books on Demand, Norderstedt. (162-172, 183-201).

Bossel, H. 2007: *Systems and Models – Complexity, Dynamics, Evolution, Sustainability*. Books on Demand, Norderstedt (108-123).

Z202 Oscillator with limit cycle (van der Pol)

Simulation task

As soon as nonlinear terms appear in its differential equation, the behavior of a system can be considerably different from that of a linear system. The pendulum already demonstrates this fact. Phenomena can appear now which do not exist in linear systems. One such phenomenon is the limit cycle. In this case the system does not move towards an equilibrium state with time but will constantly repeat an identical oscillation.

This phenomenon appears, for example, in the van der Pol oscillator. The system was initially used and described for the stabilization of electronic oscillations in radio tubes (triode vacuum tube). However, it also stabilizes cardiac activity and it appears in flow-induced oscillations (wind induced oscillation of bridges, aerodynamic flutter), in vehicle dynamics and in certain chemical reactions.

The van der Pol system consists essentially of a linear oscillator with two state variables x and y which, however, has been modified by a nonlinear structural addition so that amplifications of y arise for small x, and damping of y for large x. This causes the system to stabilize very quickly independently of the initial conditions in a stable oscillation of constant amplitude. This stable oscillation (limit cycle) appears as a closed curve in the state diagram (x, y).

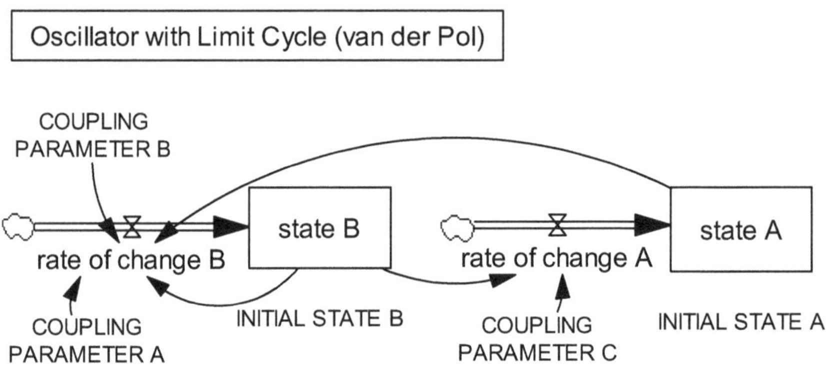

Figure Z202a: Simulation diagram for the oscillator with limit cycle.

Simulation model

The model is documented in the simulation diagram of Figure Z202a and in the following model equations. The direct mutual coupling of the state variables *state A* = x and *state B* = y corresponds basically to the harmonious undamped oscillator. The self-coupling of y (COUPLING PARAMETER A), which causes a damping if it is negative, is now modified by the current state of x in a way that causes a damping of y for positive COUPLING PARAMETER B and large x, and amplification for small x. This sign change is caused by the term $(1 - x^2)$.

The simulation model corresponds to the system of differential equations

$$dx/dt = c\,y$$
$$dy/dt = ay + b(1 - x^2)\,x$$

Parameters and initial states
COUPLING PARAMETER A = -1 [1/Second]
COUPLING PARAMETER B = 1 [1/Second]
COUPLING PARAMETER C = 1 [1/Second]
INITIAL STATE A = 0
INITIAL STATE B = 1

Dynamics
rate of change A = COUPLING PARAMETER C *state B [1/Second]
rate of change B = COUPLING PARAMETER A *state A
        +COUPLING PARAMETER B *(1 -state A^2) *state B [1/Second]
state A = INTEG (rate of change A, INITIAL STATE A) [1]
state B = INTEG (rate of change B, INITIAL STATE B) [1]

Simulation time parameters
INITIAL TIME = 0 [Second]
FINAL TIME = 20 [Second]
TIME STEP = 0.01 [Second]

Simulation results

Figure Z202b shows that for INITIAL STATE A = 0 and INITIAL STATE B = 1 an un-damped oscillation quickly develops, without the system being excited by any input function. Even if initial conditions are changed, an identical oscillation develops after short time. The state diagram of Figure Z202c shows more clearly that the system oscillates with an identical limit cycle even for very different initial conditions. The state diagram Figure Z202d (computed with auxiliary module Z115) confirms this result for 100 different initial states. The system has an unstable equilibrium point at the origin (0, 0) of the state plane. From an initial state of either within or outside the limit cycle, the motion always moves quickly towards the limit cycle and stabilizes itself there in an undamped oscillation.

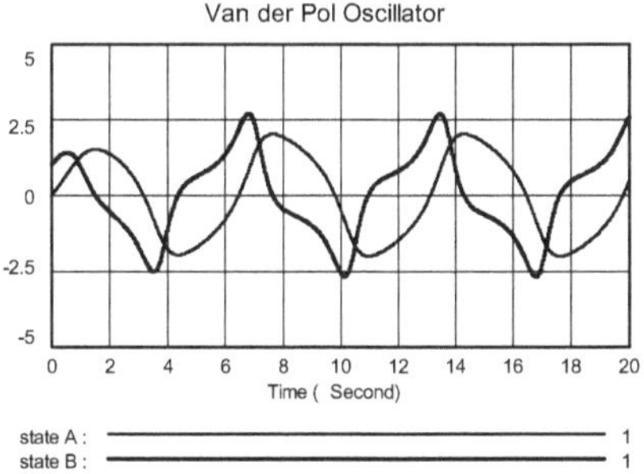

Figure Z202b: Natural oscillation of the van der Pol system.

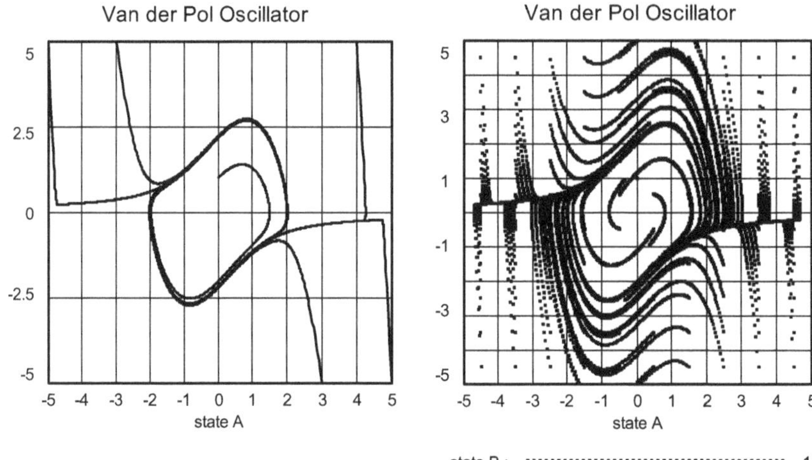

Figure Z202c: Starting from very different initial conditions, the system quickly assumes motion on its limit cycle.
Figure Z202d: Global behavior of the van der Pol system.

The COUPLING PARAMETER A $= a$ affects the frequency of the oscillation. To maintain an oscillation, a must be < 0. With $a = -1$ the system quickly approaches a stable limit cycle oscillation of period 6.6.

Exercises

1. Start by selecting COUPLING PARAMETER A $= a = -1$ and try to understand the system, the state diagram, and the time response. Identify location and type of the equilibrium point, the limit cycle, and its stability.
2. Examine the oscillation response for different values of $a < 0$.
3. Determine the state trajectories for different initial conditions within and outside the limit cycle for different frequency parameters a (COUPLING PARAMETER A).
4. Explore how system behavior, equilibrium point, and limit cycle change as COUPLING PARAMETERS A, B and B are changed. Sketch and describe some particularly remarkable results.
5. Linearize the state equations at the equilibrium point and determine the system matrix coefficients of the linearized system at this point. Insert the coefficients into model Z114 "Linear oscillator" and examine system behavior by producing the state trajectory map (using module Z115). Does the behavior of the linearized system agree with the van der Pol behavior at the equilibrium point?
6. Find the eigenvalues of the system linearized at the equilibrium point. Where are they located in the complex number plane, and what conclusions concerning stability and oscillation follow from this? (These are only valid in the neighborhood of the equilibrium point!)

References

Beltrami, E. 1987: *Mathematics for Dynamic Modeling*. Academic Press, Orlando FL (182-189)
Csaki, F. 1972: *Modern Control Theories – Nonlinear, Optimal and Adaptive Systems*. Akademiai Kiado, Budapest (359-362).
Guckenheimer, J., Holmes, P. 1983/86: *Nonlinear Oscillations, Dynamical Systems, and Bifurcations of Vector Fields*. Springer, Heidelberg and New York (67-82).
Thompson, J. M. T., Stewart, H. B. 1986/1991: *Nonlinear Dynamics and Chaos – Geometrical Methods for Engineers and Scientists*. John Wiley, Chichester UK and New York.

Z203 Brusselator

Simulation task

The kinetics of chemical reactions can be expressed by nonlinear differential equations. Because of these nonlinearities, reacting chemical systems can show a variety of phenomena. A chemical system studied particularly well is the Belousov-Zhabotinsky reaction, which exhibits complex temporal and spatial oscillations and even chaotic behavior. A simplified model of this reaction is the so-called Brusselator. It has the differential equations

$$dx/dt = A - (B+1)x + x^2 y$$

$$dy/dt = Bx - x^2 y$$

in which the parameters A and B represent constant predefined concentrations.

Simulation model

Figure Z203a shows the simulation diagram; the model equations are listed in the following.

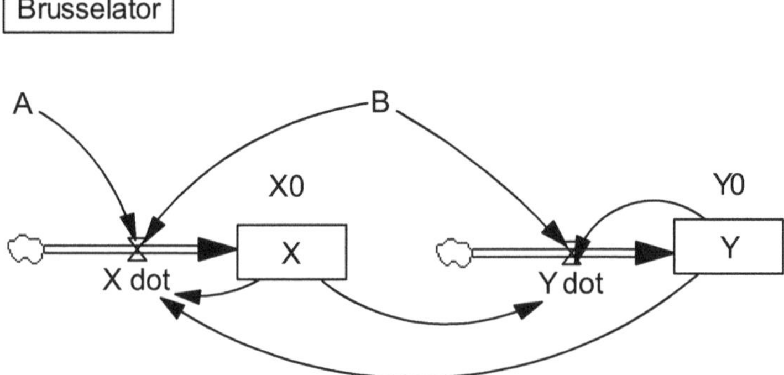

Figure Z203a: Simulation diagram of the Brusselator.

Parameters and initial values
A = 1
B = 3
X0 = 1
Y0 = 4

Dynamics
X dot = A -(B+1) *X +X^2*Y
Y dot = B*X -X^2*Y
X = INTEG (X dot, X0)
Y = INTEG (Y dot, Y0)

Simulation time parameters
INITIAL TIME = 0 [Second]
FINAL TIME = 50 [Second]
TIME STEP = 0.001 [Second]
SAVEPER = 0.01 [Second]

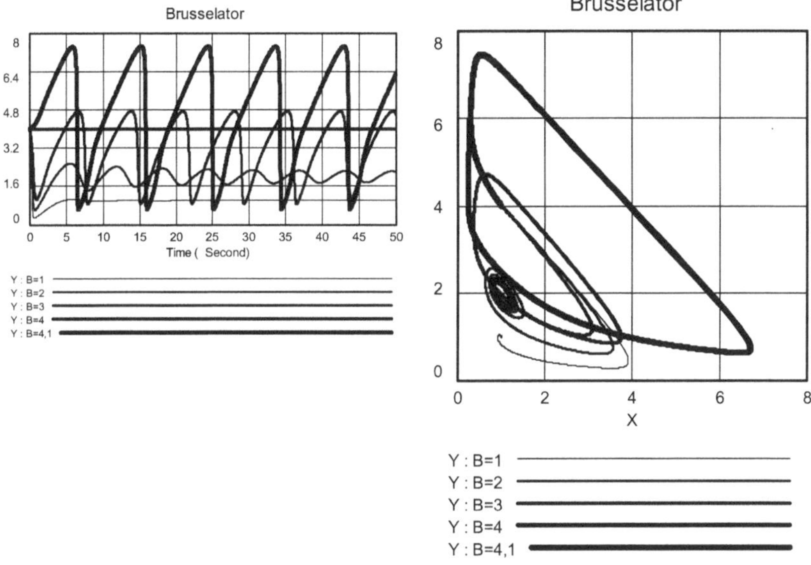

Figure Z203b: Natural oscillations as a function of parameter *B*.
Figure Z203c: Limit cycle or damped oscillation as function of parameter *B*.

Simulation results

With the default settings and different values for parameter *B* distinctive oscillations of different frequency appear in the time diagram of Figure Z203b. If the results from the same runs are drawn in a state diagram (Figure Z203c), then the following can be observed: For $B = 1$ and 2 the system oscillates in damped motion around an equilibrium point where it comes to rest after some time. For $B = 3$ the system quickly generates a limit cycle. For $B = 4$ the system remains in equilibrium in its initial state ($X0 = 1$, $Y0 = 4$). If *B* is increased slightly, then the system again oscillates in a limit cycle.

The behavior can be better understood by analyzing the system of differential equations. If the differentials dx/dt and dy/dt are set to zero it becomes obvious that the system has an equilibrium point for $x^* = A$ and $y^* = B/A$. Further, it can be shown that the equilibrium point is a stable focus for $B < A^2 + 1$, turning into an unstable focus if $B > A^2 + 1$. The critical value $B_c = A^2 + 1$ characterizes a Hopf bifurcation, where (for small *B*) the stable equilibrium point develops into a limit cycle containing an unstable equilibrium point in its interior. For $A = 1$ (default setting) the critical value is $B_c = 2$. For the parameter settings chosen for this investigation, the equilibrium points are located at $x^* = A = 1$ and $y^* = B/A = 1, 2, 3, 4$, and 4.1. A limit cycle appears for $B > 2$. For $B = 4$ the initial state coincides with the equilibrium point within the limit cycle, and the state therefore does not change.

Comparison of the state diagrams for $B = 1.5$ and $B = 2.5$ in Figures Z203d and e clearly shows the different behavior on the two sides of the critical value B_c. For $B = 1.5$ a stable focus appears; for $B = 2.5$ a limit cycle establishes itself.

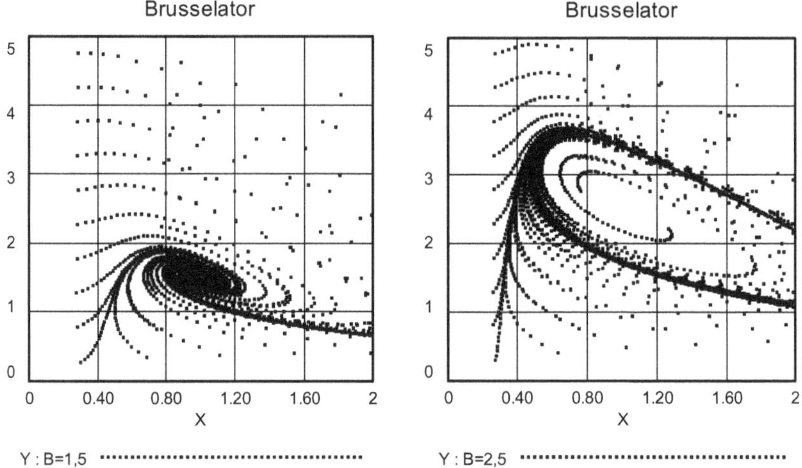

Figure Z203d: Stable focus for $B < 2$.
Figure Z203e: Limit cycle for $B > 2$.

Exercises

1. Examine the behavior as function of parameters A and B. Confirm the critical value B_c where the damped oscillation changes into a limit cycle.
2. Couple the computation of the state diagram (module Z115) to the Brusselator und generate state trajectory maps (similar to Figures Z203d and e) for interesting system behaviors.
3. Obtain background information for the Brusselator and the chemical reaction it represents (see references). Translate the results of simulation and mathematical analysis into a description of the concrete chemical process and its dynamics.

References

Nicolis, G., Prigogine, I. 1989: *Exploring Complexity – An Introduction*. W. H. Freeman, New York.
Thompson, J. M. T., Stewart, H. B. 1986/1991: *Nonlinear Dynamics and Chaos – Geometrical Methods for Engineers and Scientists*. John Wiley, Chichester UK and New York (56-60).

Z204 Bistable oscillator

Simulation task

A "small" modification of the differential equations of a linear system by a nonlinear term can cause a qualitative change of the dynamics and introduce a surprising variety of behaviors. Behaviors emerge of which linear systems are not capable – like limit cycles, or behavior determined by several equilibrium points, or even chaotic behavior. In physics and engineering, systems with a Duffing term – the appearance of a state variable in the third power – are of considerable practical importance, for example for elastic structures excited by oscillations.

In its simplest form such a system consists at its core of a linear oscillator with an additional negative coupling from x^3 to y. Depending on initial conditions, the system will now move to one of two stable equilibrium points during a damped oscillation process. The system can be physically realized by a leaf spring which is suspended between two permanent magnets. Depending on initial conditions the vibrating motion of the leaf spring ends at one of the two permanent magnets, i.e. the initial condition determines the eventual equilibrium state. Such a system therefore has the qualities of a flip flop circuit, and can be used to sort initial states into one of two categories.

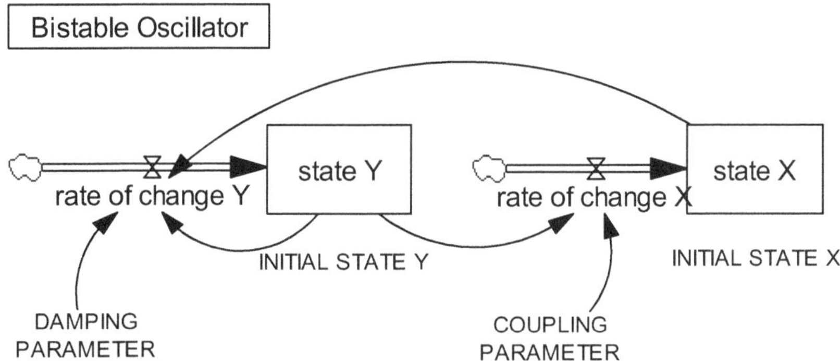

Figure Z204a: Simulation diagram of the bistable oscillator.

Simulation model

The model is documented in the simulation diagram of Figure Z204a and in the following model equations. Its basic structure is that of the linear oscillator with damping corresponding to DAMPING PARAMETER = d acting on *state Y* = y. The *rate of change X* of the other state variable *state X* is proportional to *state Y* and the COUPLING PARAMETER = b as in a linear system. In addition there is now also a restoring force proportional to *state X*, which is typical, for example, for the buckling of beams.

The system of differential equations of this model is given by

$$dx / dt = b\, y$$
$$dy / dt = x - x^3 - d\, y$$

Parameters and initial states
d = DAMPING PARAMETER = 1 [1/Second]
b = COUPLING PARAMETER = 1 [1/(Second*Second)]
INITIAL STATE X = 0 [1]
INITIAL STATE Y = 2 [Second]

Dynamics
rate of change X = state Y *COUPLING PARAMETER [1/Second]
rate of change Y = state X -DAMPING PARAMETER *state Y -state X^3 [1]
state X = INTEG (rate of change X, INITIAL STATE X) [1]
state Y = INTEG (rate of change Y, INITIAL STATE Y) [Second]

Simulation time parameters
INITIAL TIME = 0 [Second]
FINAL TIME = 20 [Second]
TIME STEP = 0.01 [Second]

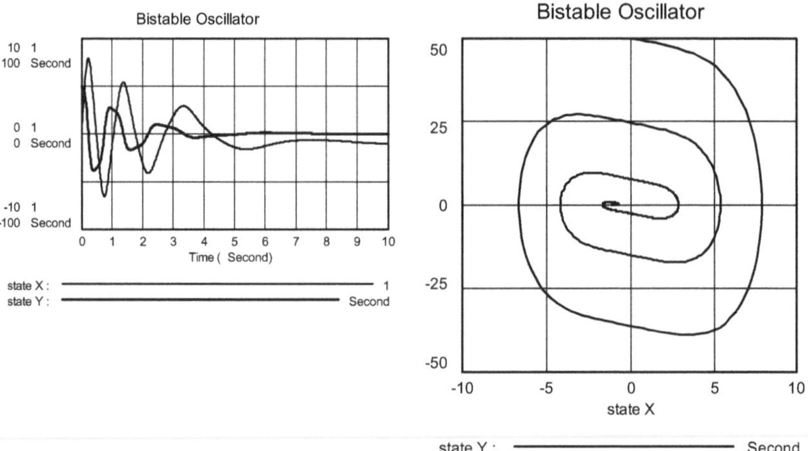

Figure Z204b: Damped oscillation ending at one of two equilibrium points.
Figure Z204c: State diagram of the bistable oscillation.

Simulation results

Figure Z204b shows the time response for the default parameters setting $(d = 1)$ and an INITIAL STATE Y of 50. A strongly damped oscillation appears. This becomes even more evident in the state diagram of Figure Z204c. In both diagrams it can be recognized that the system moves towards an equilibrium point at $x^* = -1$, $y^* = 0$. If the state trajectories are drawn for different initial states (as in Figure Z204d), it becomes obvious that some of the trajectories are "caught" by the equilibrium point at $x^* = -1$, $y^* = 0$, others by a further equilibrium point at $x^* = 1$, $y^* = 0$. Initially, the state trajectories appear to be attracted by the (unstable) equilibrium point $(0, 0)$, but are repelled by it before they can reach it, and finally end at one of the two stable equilibrium points.

Introducing the equilibrium condition $dx/dt = dy/dt = 0$ into the state equations yields two stable equilibrium points at $x = 1$ and $x = -1$, and an unstable one at $x = 0$. The free motion runs towards one of the stable equilibrium points; the place of rest depends on the initial conditions. The damping of the state variable y is dependent on y itself (velocity) and on the DAMPING PARAMETER. From Figure Z204d it can be implied that there must be a separating surface (separatrix) between trajectories leading to the two stable equilibrium points. Just like the trajectories, it is wrapped around the equilibrium points.

The global behavior in state space becomes more obvious in the state diagram of Figure Z204e (generated by coupling model Z205 to module Z115): All state trajectories are sorted to end either at the right or at the left equilibrium point. Depending on damping, very different state trajectory maps are obtained. The initial state always has a defining influence on the final state.

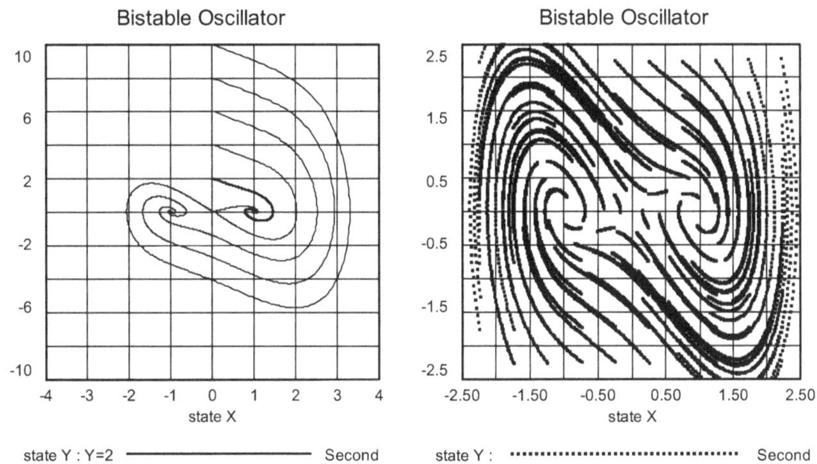

Figure Z204d: Depending on initial condition, state trajectories end either at the left or the right hand equilibrium point.
Figure Z204e: Global behavior of the bistable oscillator near the three equilibrium points (two stable foci, one unstable saddle).

Exercises

1. Examine which time response is generated by different DAMPING PARAMETERS and INITIAL STATES.
2. Couple module Z115 "State space diagram" to the model and examine the state trajectory maps for different DAMPING PARAMETER $0 < d < 1$.

3. Enter the parameter values for the linear system (without the x^3 term) in model Z114 "Linear oscillator" and compare the resulting state diagram with that for model Z204.

4. Linearize the system at the equilibrium point ($x = 0$, $y = 0$). What are the corresponding values of the system parameters, and how do they differ from the linear system obtained by dropping the x^3 term?

5. Compute the state diagram of the system for different COUPLING PARAMETERS b and DAMPING PARAMETERS d. Discuss system behavior by reference to the state diagrams. How do behavior and stability change?

6. Linearize the nonlinear state equations at the two equilibrium points $y^* = 0$, $x^* = \pm 1$, and investigate system behavior at these equilibrium points using the substitute linear system (model Z114).

7. Find the eigenvalues of the linearized system at these points. Compare the behavior and stability conclusions of the original and the linearized system (state diagram and root locus) at these points (concerning linearization, eigenvalues, root locus etc. see e.g. Bossel 1994 Ch. 7, Bossel 2007:124-147).

References

Bossel, H. 1994: *Modeling and Simulation*. A K Peters, Wellesley MA.

Bossel, H. 2007: *Systems and Models – Complexity, Dynamics, Evolution, Sustainability*. Books on Demand, Norderstedt.

DeRusso, P. M., Roy, R. J., Close, C. M. 1965: *State Variables for Engineers*. Wiley, New York (483-488).

Guckenheimer, J., Holmes, P. 1983/86: *Nonlinear Oscillations, Dynamical Systems, and Bifurcations of Vector Fields*. Springer, Heidelberg and New York (S. 82-91).

Thompson, J. M. T., Stewart, H. B. 1986/1991: *Nonlinear Dynamics and Chaos – Geometrical Methods for Engineers and Scientists*. J. Wiley, Chichester / New York.

Z205 Chaotic bistable oscillator (Duffing)

Simulation task

Intuitively one would expect a system with such regular behavior as the bistable oscillator (model Z204) to behave in a similar regular and predictable fashion if subjected to periodic excitation. Surprisingly, this is not the case. In a wide parameter range the system exhibits an astonishing variety of behaviors, depending on amplitude of the periodic forcing and the damping in the system (cf. Figure 1.10 in Thompson and Stewart 1986). For certain parameter combinations chaotic behavior arises.

This fact is of great importance in some branches of technology, in particular where elastic deformations combine with periodic forcing (as in the case of airplane wings). The process can be physically realized by a leaf spring fixed in a frame such that the free end is suspended over two permanent magnets (cf. Guckenheimer and Holmes 1983: 82-91). The free end of the leaf spring will not remain centered between the magnets, but will bend to one of the magnets after a small perturbation, and remain there. If released from an initial large displacement, it will oscillate back and forth between the magnets, and finally come to rest at one of them. This process was described in model Z204 "Bistable oscillator".

If the frame containing leaf spring and magnets is moved back and forth in periodic oscillation, the behavior of the system changes drastically. The present model is developed to simulate this behavior.

Simulation model

The simulation diagram of Figure Z205a and the following model equations document the model. It corresponds exactly to model Z204 "Bistable oscillator", except for the sinusoidal periodic forcing with frequency ω and amplitude factor q. Chaotic motion now arises for a wide parameter range (not for all parameter constellations!). The basic structure is again that of a damped linear oscillator subject to sinusoidal forcing. In addition, the system includes the nonlinear restoring force (proportional to x^3). The system of differential equations is now:

$$dx/dt = a\,y$$
$$dy/dt = -d\,y + x - x^3 + q\cos\omega t$$

Parameters and initial states
d = DAMPING PARAMETER = 0.25 [1/Second]
a = COUPLING PARAMETER = 1 [1/(Second*Second)]
q = AMPLITUDE = 0.3 [1]
ω = FREQUENCY = 1 [1/Second]
INITIAL STATE X = 0 [1]
INITIAL STATE Y = 1 [Second]

Dynamics
excitation = AMPLITUDE *COS (FREQUENCY *Time) [1]
rate of change X = state Y *COUPLING PARAMETER [1/Second]
rate of change Y = state X -DAMPING PARAMETER *state Y -state X^3 +excitation
    [1]

state X = INTEG (rate of change X, INITIAL STATE X) [1]
state Y = INTEG (rate of change Y, INITIAL STATE Y) [Second]

Simulation time parameters
INITIAL TIME = 0 [Second]
FINAL TIME = 100 [Second]
TIME STEP = 0.01 [Second]

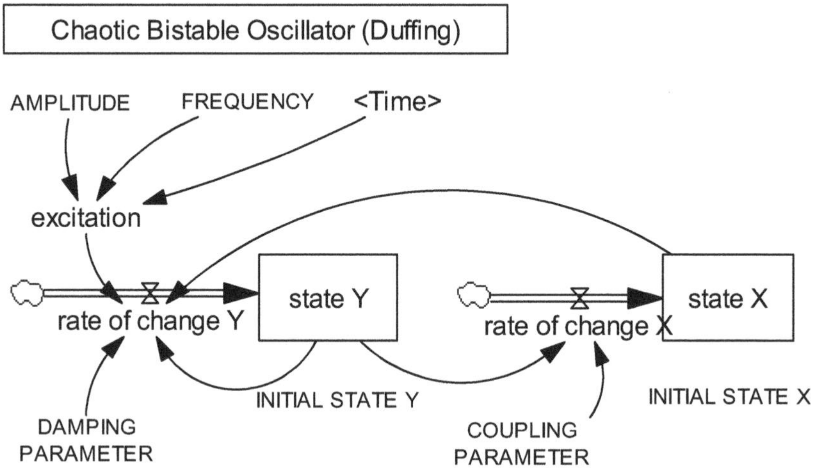

Chaotic Bistable Oscillator (Duffing)

AMPLITUDE   FREQUENCY   <Time>

excitation

rate of change Y   state Y   rate of change X   state X

INITIAL STATE Y   INITIAL STATE X

DAMPING PARAMETER   COUPLING PARAMETER

Figure Z205a: Simulation diagram for the chaotic bistable oscillator.

Simulation results

Initial simulation runs reveal that the results depend strongly on integration method and its step width (TIME STEP). The Runge-Kutta procedure of 4th order (RK4) with the step width of the default setting is therefore employed in the following. The results are then no longer significantly affected by the choice of step width. Figure Z205b presents the time diagram of a simulation with the default parameter setting. We obtain apparently completely irregular behavior for *state X* and *state Y* which does not even correlate with the excitation frequency. In the state diagram of Figure Z205c the behavior becomes more obvious: The state trajectory sometimes moves around one, then around the other equilibrium point before suddenly jumping again to the other range of attraction.

A small change of parameters can cause a significant change of behavior. Figures Z205d and Z205e show chaotic behavior at relatively high values for AMPLITUDE (= 1.7) and FREQUENCY (= 2.3). Between chaotic regions there are other regions where the system oscillates rather regularly around one or the other equilibrium point or around both of them. By changing the three parameters FREQUENCY ω, AMPLITUDE q, DAMPING PARAMETER d a great variety of behaviors can be produced. An overview is best obtained by interactive simulation where parameters can be changed between simulations (by slide bars), and small time diagrams of all variables are simultaneously presented in the simulation diagram on the computer screen (option *SyntheSim* in VenSim/VenPLE).

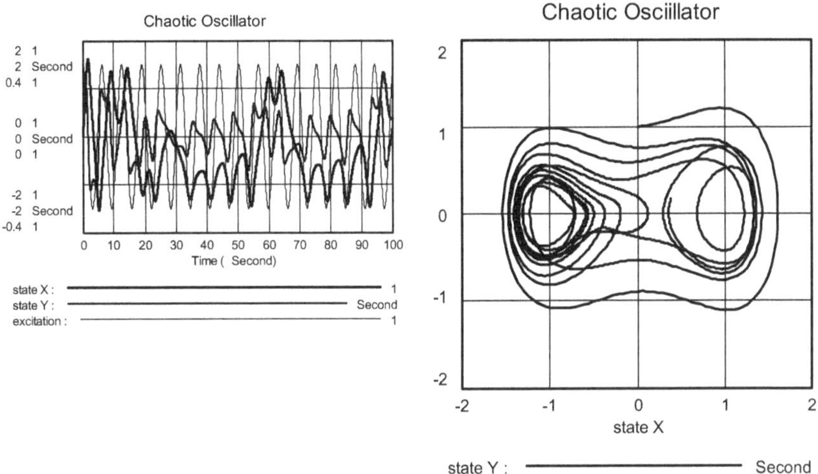

Figure Z205b: Chaotic behavior is evident in the time diagram (default parameter setting).
Figure Z205c: State space trajectory of chaotic behavior.

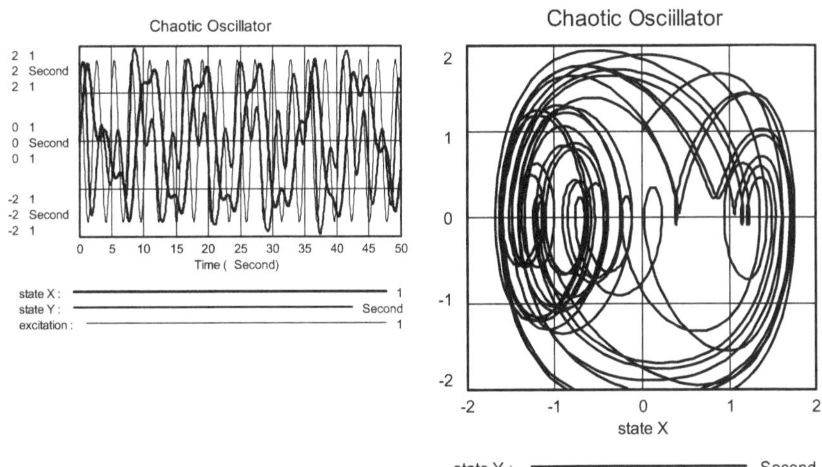

Figure Z205d: Chaotic behavior at higher values of frequency and amplitude of the periodic forcing.
Figure Z205e: The same process in the state diagram.

Exercises

1. Investigate the behavior for different values of the three parameters FREQUENCY ω (in the range of about 0 to 3), AMPLITUDE q (0 to 2.5), DAMPING PARAMETER d (0 to 1) using time plots and state diagrams (x, y). (Preliminary examination using *Synthe-Sim*; individual simulations for documenting interesting cases).
2. Investigate (using FREQUENCY = const = 1, i.e. default value) the dependence of behavior modes on AMPLITUDE q (0 to 2.5) and DAMPING PARAMETER d (0 to 1), and draw a diagram (d over q) showing the different behavior ranges.
3. Leave q and d at their default settings, change FREQUENCY ω (in the range of about 0 to 3), and determine for the different cases the dominant frequency in the time plot for *state X*. Plot this frequency as function of excitation FREQUENCY ω. Describe and discuss the results.
4. Apply minimally different values for the INITIAL STATE Y (1, 1.01, 1,001, 1.0001) and compute and compare the simulation results for e.g. *state X*. (*Note*: In the Ven-Sim/VenPLE simulation software you can store different runs (and their results) under different names and then select a quantity by clicking on it in the simulation diagram and selecting "graph" (on the left) to plot it in a common diagram.) What do you find? Repeat the same experiment with model Z204 "Bistable oscillator". Referring to these observations, explain the mechanism causing chaotic dynamics.

References

Guckenheimer, J., Holmes, P. 1983/86: *Nonlinear Oscillations, Dynamical Systems, and Bifurcations of Vector Fields*. Springer, Heidelberg and New York (82-91).
Thompson, J. M. T., Stewart, H. B. 1986/1991: *Nonlinear Dynamics and Chaos – Geometrical Methods for Engineers and Scientists*. John Wiley, Chichester UK and New York (1-107).

Z206 Heat, weather, and chaos (Lorenz system)

Simulation task

Weather forecasts are an important application of computer simulation. During his investigation of the nonlinear partial differential equations describing the coupling of convection and thermal conduction in the atmosphere the meteorologist Lorenz 1963 stumbled upon a phenomenon which has fascinated scientists ever since: chaos. Lorenz had tried to solve these differential equations by a Fourier series approach. He noticed that only three terms have relatively major importance: one (x) describing the velocity profile, and two (y and z) describing the temperature distribution.

For these state variables, Lorenz formulated the following approximate system of (nondimensional) differential equations:

$$dx/dt = a(y - x)$$
$$dy/dt = b x - y - x z$$
$$dz/dt = -c z + x y$$

The (dimensionless) parameters a and b are hydrodynamic parameters (the Prandtl number and the Rayleigh number), the parameter c is a geometry parameter. The Prandtl number expresses the relationship between thermal conduction and heat convection in the fluid. The Rayleigh number corresponds approximately to the temperature difference between upper atmosphere and the earth's surface in this particular application.

When attempting to solve this system numerically, Lorenz found completely different time behavior of the system for minute changes in initial conditions. This phenomenon originates in the physical relationships and is not caused by problems with numerical integration. It therefore also determines actual weather development and causes weather forecasts for periods of several days to become extremely uncertain even with the best computer models. In a pointed formulation of this mathematical dilemma, the proverbial "stroke of a butterfly wing" in the Niger delta can (theoretically) lead to the formation of a tropical depression developing into a hurricane while crossing the Atlantic, causing great damage in Florida a few days later.

Lorenz' mathematical model produces chaotic behavior only in certain parameter ranges. The model generally describes fluid flows over a heated surface, where cellular flows can develop under certain conditions (Bénard cells). These are actually observed in meteorology, on the surface of the sun, in flat ponds, and in fluid flows with cooled upper surfaces. Other chaotic systems (laser, dynamos: model Z208 "Coupled dynamos and chaos") show a similar basic system structure.

Simulation model

The simulation diagram of Figure Z206a and the following program statements document the model. The default parameter settings are the system parameters usually quoted in the literature.

The model is distinguished by two nonlinearities ($x \cdot z$, $x \cdot y$) and by the possible sign reversal of the rate $dx/dt = a (y - x)$. Viewed in isolation, these characteristics represent nothing unusual; the chaotic behavior arises only from the complete structure of the model. In general it is true that chaotic behavior cannot be recognized by looking at system structure.

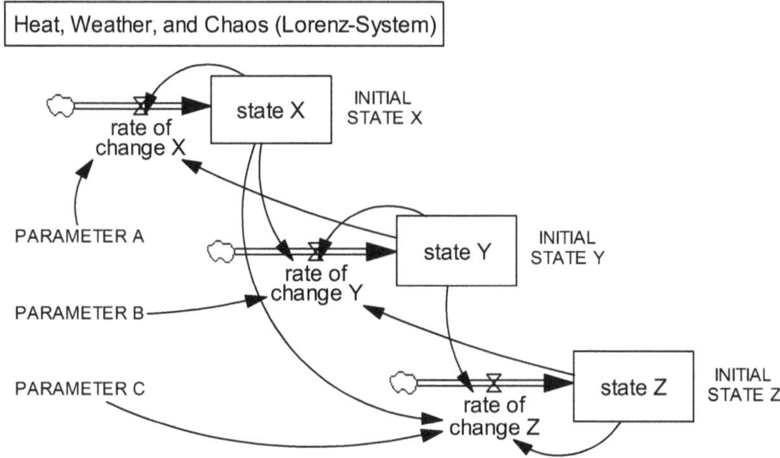

Figure Z206a: Simulation diagram for the Lorenz system.

Parameters and initial states
PARAMETER A = 10
PARAMETER B = 28
PARAMETER C = 2.667
INITIAL STATE X = 15
INITIAL STATE Y = 15
INITIAL STATE Z = 15

Dynamics
rate of change X = PARAMETER A *(state Y -state X)
rate of change Y = PARAMETER B *state X -state Y -state X *state Z
rate of change Z = -PARAMETER C *state Z + state X *state Y
state X = INTEG (rate of change X, INITIAL STATE X)
state Y = INTEG (rate of change Y, INITIAL STATE Y)
state Z = INTEG (rate of change Z, INITIAL STATE Z)

Simulation time parameters
INITIAL TIME = 0
FINAL TIME = 20
TIME STEP = 0.01

Simulation results

In the time plots of the three state variables (Figure Z206b) one recognizes an irregular sequence of amplifying oscillations and abrupt changes of state. Note that in contrast to model Z205 "Bistable oscillator" there is no forcing of the system, let alone periodic excitation. The oscillations appearing in the time plot represent purely internally produced dynamics.

The state diagrams (Figures Z206c, d, e) reveal rather regular motion in two regions of attraction of the state space, which is not obvious in the irregular time plots. The state trajectory orbits one of two centers of attraction, and then flips to the other region of attraction after an unforeseeable number of cycles. From this a typical butterfly shape of the attractor arises. A three-dimensional representation (x, y, z) (which is not possible with the program systems used here) would show the shape of the attractors in full state space. The state diagrams shown here are projections of three-dimensional state space on the two-dimensional phase planes (x, z), (y, z), (x, y).

Note that state trajectories can never coincide at a given point in state space since otherwise the deterministic motion would repeat itself (i.e. the states, rates, and hence the motion would be identical, see the state equations). Three-dimensional (or higher dimensional) state trajectories can only cross in a two-dimensional (or lower dimensional) projection, never in three-dimensional (or higher dimensional) state space. If trajectories (of three state variables) are viewed from different angles in three-dimensional state space, one can verify that they never cross or coincide.

The state trajectory map of this attractor and the behavior of the system change with parameter choice. A critical parameter for the appearance of chaos is in particular PARAMETER B (here $b = 28$).

Exercises

1. Repeat the simulations with the default parameter settings but minimal changes (e.g. third place behind the decimal point) of INITIAL STATE for a state variable. Draw the different time plots for a particular state variable in one diagram (cf. the suggestions for exercises for model Z205). Compare and discuss the results.
2. Change PARAMETER B = b and determine the range of b for which chaos appears.
3. Examine the influence of the other parameters on system behavior, particularly on the appearance of chaos.
4. Transfer the simulation results for the time series of the three state variables to a corresponding program (write your own program, if necessary) and draw the attractor in a three-dimensional representation. Adjust the view angle to obtain impressive pictures of the "butterfly".
5. Generate three-dimensional representations of other interesting attractor shapes which result from changes of parameters.

References

Canty, M. J. 1995: *Chaos und Systeme – Eine Einführung in Theorie und Simulation dynamischer Systeme*. Vieweg, Braunschweig und Wiesbaden (112-132).
Jetschke, G. 1989: *Mathematik der Selbstorganisation – Qualitative Theorie nichtlinearer dynamischer Systeme und gleichgewichtsferner Strukturen in Physik, Chemie und Biologie.* VEB Deutscher Verlag der Wissenschaften, Berlin (130-136).
Guckenheimer, J., Holmes, P. 1983/86: *Nonlinear Oscillations, Dynamical Systems, and Bifurcations of Vector Fields*. Springer, Heidelberg and New York (92-102).
Thompson, J. M. T., Stewart, H. B. 1986/1991: *Nonlinear Dynamics and Chaos – Geometrical Methods for Engineers and Scientists*. John Wiley, Chichester UK and New York (212-234).

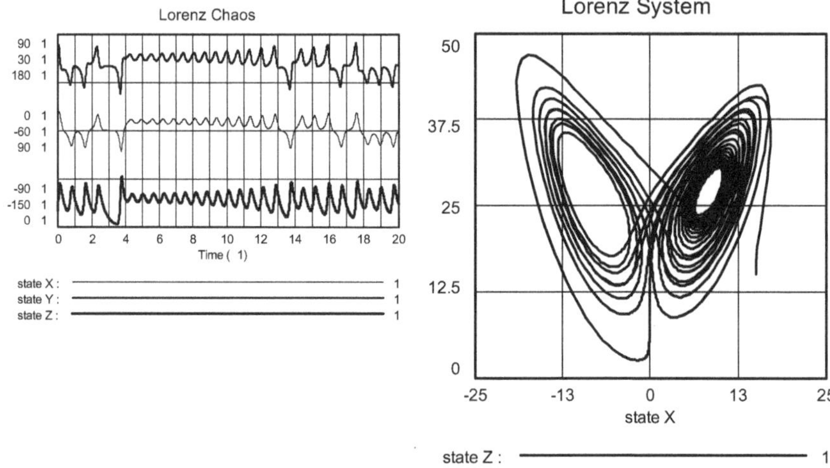

Figure Z206b: Chaotic behavior in the time plot for the Lorenz system.
Figure Z206c: Projection of the state trajectory in the (x, z) plane.

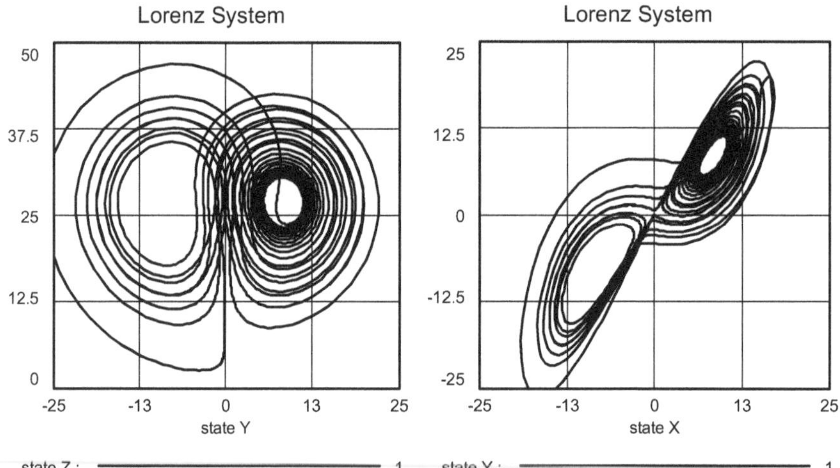

Figure Z206d: Projection of the state trajectory in the (y, z) plane.
Figure Z206e: Projection of the state trajectory in the (x, y) plane.

Z207 Chaotic attractor (Rössler)

Simulation task

To better understand the qualities of chaotic attractors, the mathematician Rössler developed (1976) a system which can generate chaotic behavior for certain parameter ranges of a single nonlinear term. The attractor is also described as a simply folded band. The system is a mathematical artifact and does not have any counterpart in reality but it is well suited to study the emergence of chaos and its qualities.

The Rössler attractor is defined by the differential equations

$$dx\,/\,dt = -y - z$$
$$dy\,/\,dt = x + a\,y$$
$$dz\,/\,dt = b + z\,(x - c)$$

The system – consisting at its core of a linear oscillator and a nonlinear structural modification $z\,x$ – produces a stable limit cycle with increasing period doublings for increasing parameter a (b and c are constant). As a is increased further, chaotic behavior results, which from time to time is interrupted by "windows" exhibiting regular limit cycles.

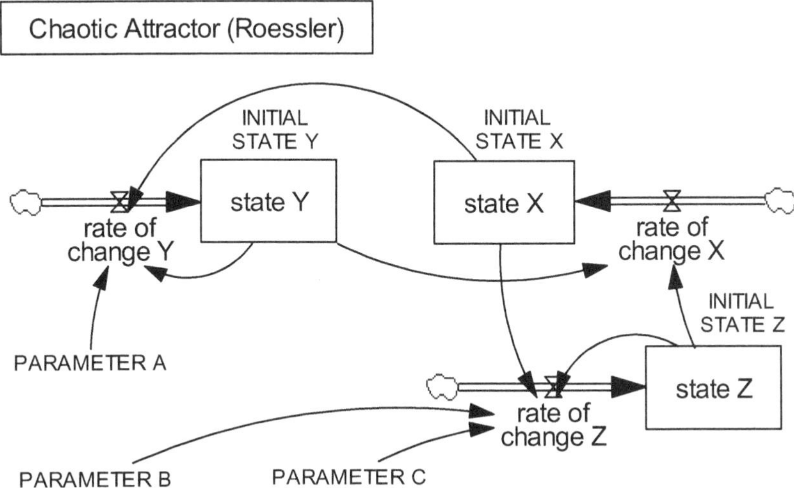

Figure Z207a: Simulation diagram for the Rössler attractor.

Simulation model

The simulation diagram of Figure Z207a and the following program statements document the model. The connections between *state X* and *state Y*, if looked at in isolation, correspond to the linear oscillator. However, the *state Z* also changes *state X*. The *rate of change Z* is determined by the difference of (*state X* – PARAMETER C) (i.e. $x - c$) which can lead to sign change of the *rate of change Z* (and cause a limit cycle).

Parameters and initial states
PARAMETER A = 0.55
PARAMETER B = 2
PARAMETER C = 4
INITIAL STATE X = 1
INITIAL STATE Y = 0
INITIAL STATE Z = 0

Dynamics
rate of change X = -state Y -state Z
rate of change Y = state X +PARAMETER A *state Y
rate of change Z = PARAMETER B +(state X -PARAMETER C) *state Z
state X = INTEG (rate of change X, INITIAL STATE X)
state Y = INTEG (rate of change Y, INITIAL STATE Y)
state Z = INTEG (rate of change Z, INITIAL STATE Z)

Simulation time parameters
INITIAL TIME = 0
FINAL TIME = 100
TIME STEP = 0.01
SAVEPER = 0.05

Simulation results

For the default parameter setting, chaotic behavior appears in the time plot (Figure Z207b). The complex shape of the attractor can be assessed from the projections of the three-dimensional chaotic attractor in the three state planes (x, z), (y, z), (x, y) (shown in Figure Z207c, d, e). To investigate the peculiarities of the system we fix PARAMETER A = a = 0.2 and PARAMETER B = b = 0.2 and vary PARAMETER C between 2 and 5. The initial values are INITIAL STATE X = 0, INITIAL STATE Y = 2, INITIAL STATE Z = 0. We observe the following (Figure Z207f through i): For PARAMETER C = c = 2 the system quickly moves from its initial state to a permanent limit cycle (with regular oscillations in the time plot).

The limit cycle is to be expected because of the sign inversion of the rate z (x – c). For c = 3 a limit cycle also develops, but the trajectory closes only after two cycles ("folded eight"), repeating the motion thereafter. The period of the limit cycle therefore has doubled. For c = 4 the period of the limit cycle doubles once again: it now takes four cycles to close the trajectory. This permanent period doubling with increasing PARAMETER C finally leads to chaotic motion (at approximately c = 4.2), which is also obvious in Figure Z207i for c = 5.

Period doubling is evident also in the time plot for *state Z* for the four runs (Figure Z207k): Regular peaks indicating a doubling of the period are observed for c = 3; for c = 4 they indicate a quadrupling of the period. In the chaotic range for c = 5 the regular periodicity disappears.

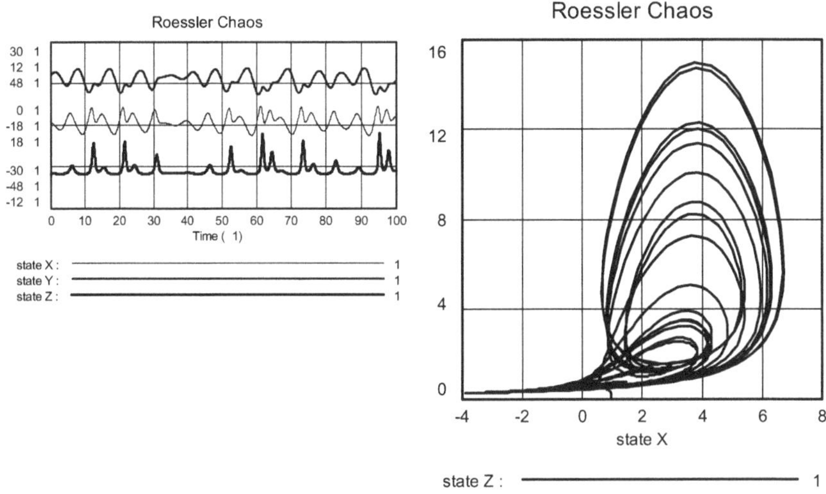

Figure Z207b: Chaotic behavior in the time diagram.
Figure Z207c: Projection of state trajectories in the (x, z) plane.

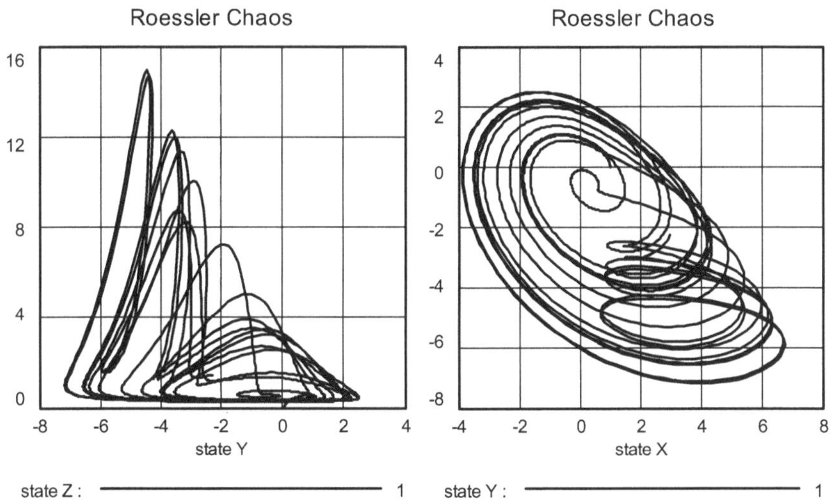

Figure Z207d: Projection of state trajectories in the (y, z) plane.
Figure Z207e: Projection of state trajectories in the (x, y) plane.

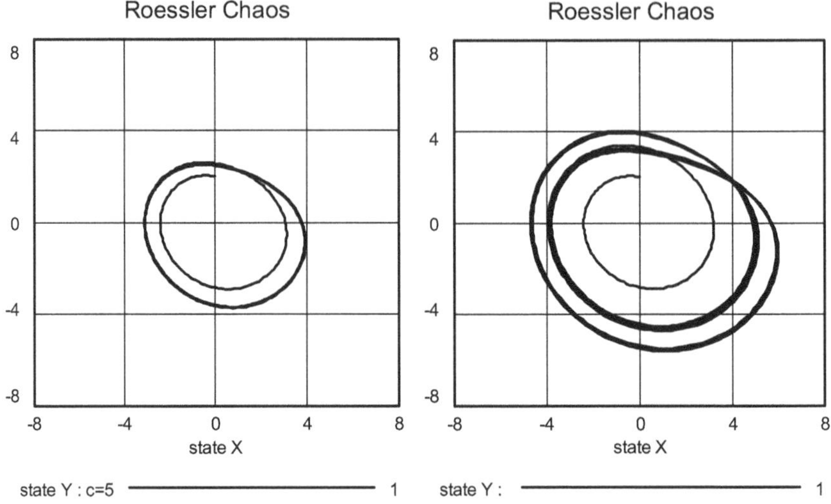

Figure Z207f: Limit cycle at low value for PARAMETER c = 2.
Figure Z207g: First period doubling for c = 3: double limit cycle.

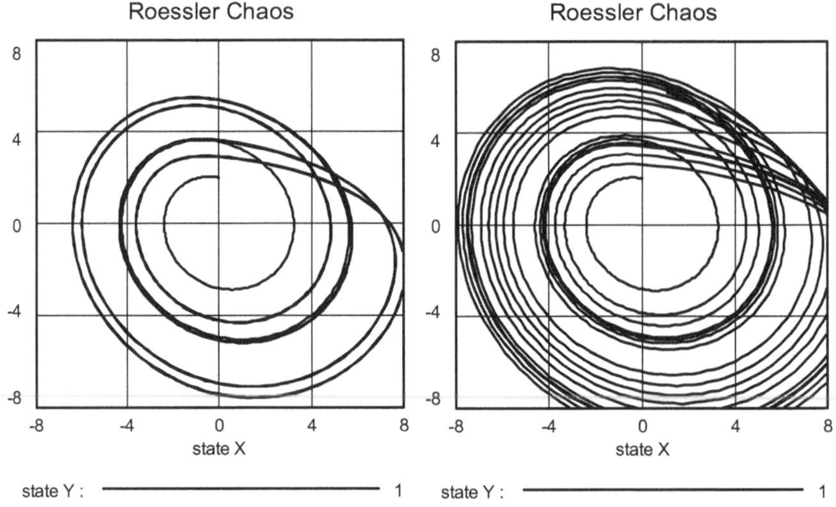

Figure Z207h: Second period duplication for c = 4: fourfold limit cycle.
Figure Z207i: Transition to chaotic behavior for c = 5.

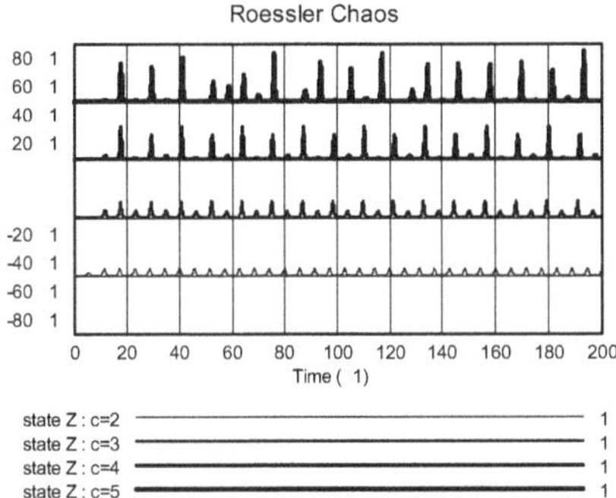

Figure Z207k: Period doubling can also be observed in the time plot.

Exercises

1. Using the default parameter setting, examine the time response and shape of the attractor as function of PARAMETER A.

2. Transfer the time series of the simulation results to a computer program for three-dimensional graphics and draw the (3D) attractor surfaces for interesting parameter combinations.

3. Draw the limit cycles and attractor surfaces for the parameters of Figure Z207f to i in three-dimensional graphics.

4. Try to relate the influence of the three parameters a, b, c to certain qualities of the dynamics (frequency, amplitude, limit cycle, appearance of chaos).

5. Retain the default parameter setting with the exception of PARAMETER A = a, but change this parameter gradually in the range of $0.3 < a < 0.5$. Document the transition from limit cycles of period 1, 2, 4 etc. to a chaotic attractor and subsequently to limit cycles of higher period.

References

Jetschke, G. 1989: *Mathematik der Selbstorganisation – Qualitative Theorie nichtlinearer dynamischer Systeme und gleichgewichtsferner Strukturen in Physik, Chemie und Biologie*. VEB Deutscher Verlag der Wissenschaften, Berlin (136-138).

Thompson, J. M. T., Stewart, H. B. 1986/1991: *Nonlinear Dynamics and Chaos – Geometrical Methods for Engineers and Scientists*. John Wiley, Chichester UK and New York (235-253).

Z208 Coupled dynamos and chaos

Simulation task

Depending on the particular linkage of components, chaos can also appear in technical systems, for example, in two coupled dynamos (Beltrami 1987: 214-218).

Let two identical dynamos be electrically coupled to each other such that the current of one dynamo excites the magnetic field of the other. The dynamos have identical self-induction L, resistance R, constant drive torque G and therefore identical angular acceleration $\omega'_1 = \omega'_2$. The latter means that the angular velocities differ only by a constant, i.e. $\omega_1 - \omega_2 =$ const. Therefore only one of these angular velocities must be introduced as state variable. The currents I_1 and I_2 of the two dynamos are also state variables. The system is therefore represented by three differential equations for the three state variables.

If dimensionless variables x and y are introduced for the currents, z for the representative angular velocity and τ for dimensionless time (for definitions see the exercises) the system can be formulated as

$$dx / d\tau = zy - \mu x$$
$$dy / d\tau = (z - \gamma)x - \mu y$$
$$dz / d\tau = 1 - xy$$

The system linkages between the state variables x and y are symmetric. The parameter γ corresponds to the (constant) difference of the angular velocities. The system has three nonlinear couplings xz, yz, xy. In the following, the parameter μ is designated as DAMPING PARAMETER A, parameter γ as COUPLING B.

Simulation model

The simulation diagram of Figure Z208a and the following program statements document the simulation model completely.

Parameters (*initial states are defined in the INTEG equations*)
μ = DAMPING PARAMETER A = 1
γ = COUPLING B = 2

Dynamics
angular acceleration = (1 -current in circuit 1 *current in circuit 2)
angular velocity = INTEG (angular acceleration, 0)
increase current 1 = angular velocity *current in circuit 2
decrease current 1 = DAMPING PARAMETER A *current in circuit 1
current in circuit 1 = INTEG (+increase current 1 -decrease current 1, 1)
increase current 2 = (angular velocity –COUPLING B) *current in circuit 1
decrease current 2 = DAMPING PARAMETER A *current in circuit 2
current in circuit 2 = INTEG (+increase current 2 -decrease current 2, 0)

Simulation time parameters
INITIAL TIME = 0
FINAL TIME = 100
TIME STEP = 0.01

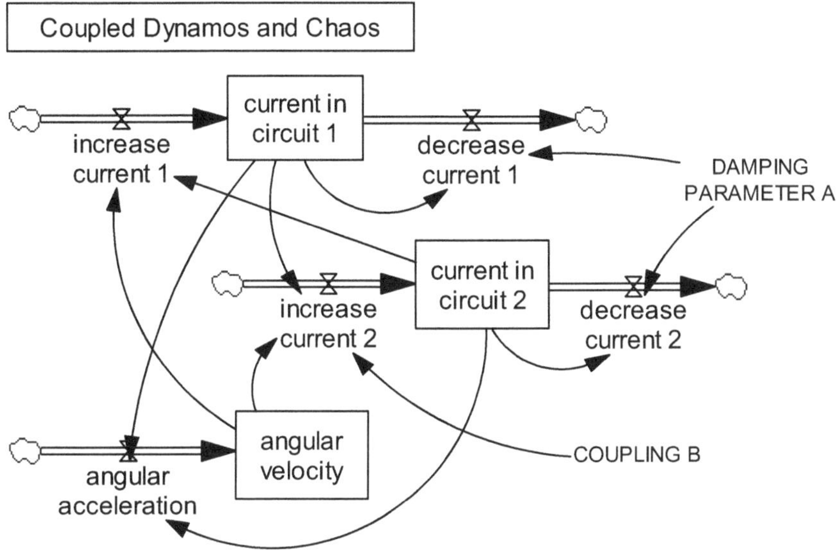

Figure Z208a: Simulation diagram for coupled dynamos.

Simulation results

The time plots of the state variables corresponding to the default parameter setting are shown in Figure Z208b. The corresponding state diagrams for angular velocity and the two currents are shown in Figures Z208c, d, and e. The state trajectory orbits one of the two unstable equilibrium points repeatedly (which means oscillations in the time response) in an unpredictable number of cycles, and then moves again to the other range of attraction. The DAMPING PARAMETER A and the COUPLING B determine the time response and the shape of the attractor.

Exercises

1. Examine the role of DAMPING PARAMETER A = μ and COUPLING B = γ on the shape of the attractor and of limit cycles using a three-dimensional representation of the attractor in state space.

2. Investigate systematically the behavior possibilities as function of damping parameter A = μ and COUPLING B = γ and sketch the regions of chaotic behavior and of limit cycles (what period?) in a diagram with coordinates μ and γ. Document the attractor surface in state space for particularly interesting cases.

3. Translate the system – formulated here with dimensionless quantities and dimensionless time – into differential equations for real quantities in real time, modify the simulation model correspondingly, check dimensional validity (use "units check" in VenSim/VenPLE), apply realistic values for the parameters, and compute time plots for currents and angular velocity.

Conversions (cf. Beltrami 1987: 214-218) with L = self-inductance, R = resistance, G = torque, I_1 and I_2 = current, ω_1 and ω_2 = angular velocity, M = mutual inductance:

$$I_i = \sqrt{G/M}\, x_i \ , \ \omega_i = \sqrt{GL/CM}\, y_i \ , \ t = \sqrt{CL/GM}\, \tau \ ,$$

$$\mu = (R/L)\sqrt{LC/MG} \ , \ \gamma = y_1 - y_2$$

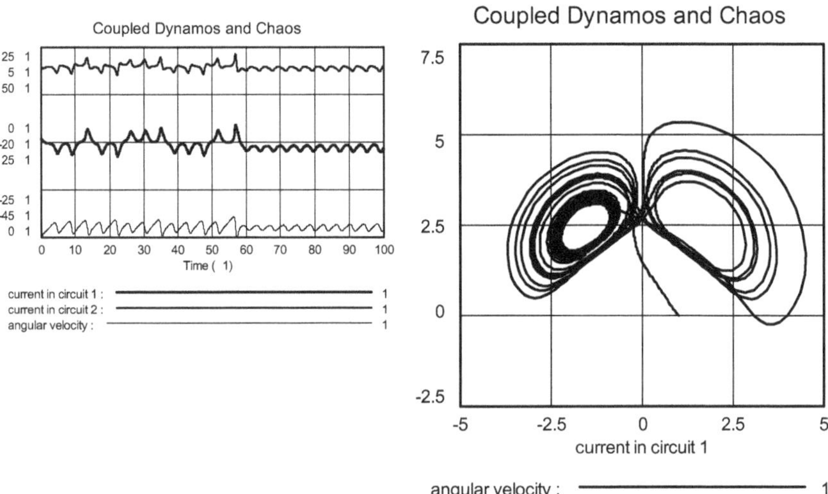

Figure Z208b: Chaotic behavior in the time plot.
Figure Z208c: Projection of the state trajectory in the (I_1, ω) plane.

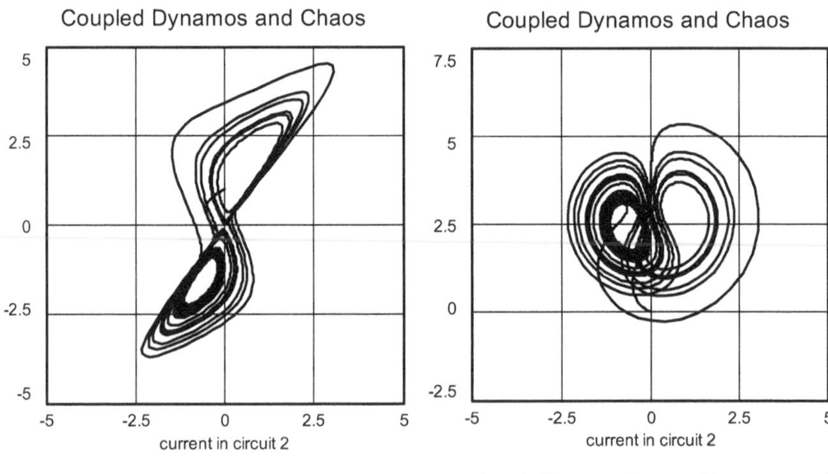

Figure Z208d: Projection of the state trajectory in the (I_2, I_1) plane.
Figure Z208e: Projection of the state trajectory in the (I_2, ω) plane.

References

Beltrami, E. 1987: *Mathematics for Dynamic Modeling*. Academic Press, Boston and New York (214-218).

Z209 Balancing an inverted pendulum

Simulation task

In principle, unstable systems can be stabilized by appropriate change of structure, or addition of stabilizing structural elements. Control theory deals with this important task of analysis and synthesis of control systems. Many technical systems would be hopelessly unstable and could not be safely used without a special structural addition of control components or a special control system. *Examples*: aircraft, rockets on take-off, chemical reactors, nuclear power stations. It is not always possible to achieve inherent stability which remains functional even if all power sources fail (as is possible for airplanes by careful balancing of the different forces and moments).

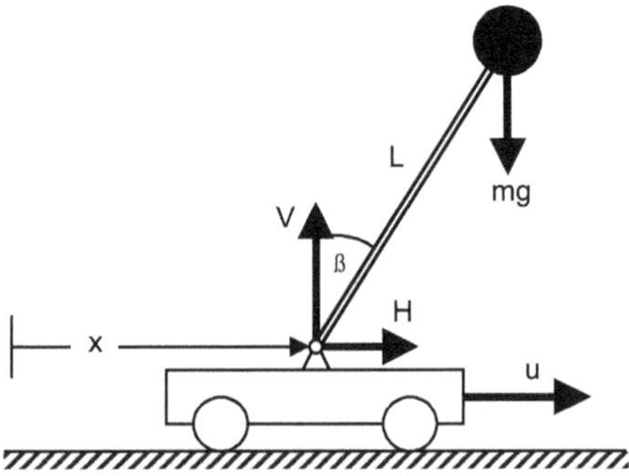

Figure Z209a: An inverted pendulum can be stabilized by a control system.

In this section we deal with the stabilization of a rigid pendulum in an upright position, which is of course hopelessly unstable unless it can be balanced by a control system that prevents it from falling over. The task corresponds to the stabilization of space rockets taking off – or of the Segway vehicle. We will start by developing a structural addition that could lead to stabilization. For this new structure the equations of motion must be formulated; they prove to be highly nonlinear. Since we assume that stabilization will succeed and pendulum excursions will remain in the neighborhood of the upper dead center (i.e. the pendulum will remain upright), we can simplify the equations of motion significantly by linearization. We then design a control function depending on the state variables of the system, leaving its parameters initially undetermined. This provides the necessary equations for the corresponding simulation model. In multiple simulations we can search for combinations of control parameters which provide good stability behavior. This search is aided by the introduction of a control efficiency criterion (minimal energy consumption). The equations of motion and control for this system have been derived elsewhere (Bossel 1992: 203-217; Bossel 1994: 234-248; Bossel 2004: 279-293); in the following the derivation is therefore presented in summarized form.

System structure and system equations

From experience (balancing a broom-stick on a finger) we know that an unstable system like the upright rigid pendulum can be stabilized at its upper dead center if the pivot is moved skillfully and rapidly such that this motion counteracts the falling motion and catches it. Therefore, a movement of the pivot is essential for stabilization, and must play a role in the controlled system and its equations of motion.

To stabilize the pendulum, we could mount its pivot on a small carriage whose wheels could be driven forward and backward by a small electric motor (Figure Z209a). (We assume that the pendulum can move only in one plane of rotation.) Sensors are mounted on the carriage and on the pendulum, providing instantaneous information about the system state to the controller directing the motion of the carriage. The complete system now consists of the pendulum and the carriage, with its drive unit controlled by a control system. Its equations of motion arise from the (usual) condition that the sums of the forces and moments acting both at the pendulum and the carriage must disappear. These differential equations are nonlinear. However, for dealing with the control task they can be linearized and considerably simplified (cf. Bossel loc. cit.). The dynamic behavior can be analyzed or simulated only after the control function has been defined.

Simulation model

With the given system equations and the adopted linear *control function* the simulation diagram can be drawn (Figure Z209b). The *moment of inertia for sphere* of the pendulum as well as the five coefficients *aa*, *a*, *b*, *c* and *d* are determined from the constant parameter values of the system (GRAVITATION CONSTANT, CARRIAGE MASS, PENDULUM LENGTH, PENDULUM MASS, RADIUS OF PENDULUM MASS). These quantities must be determined only once at the beginning of the simulation. They appear as coefficients in the simulation diagram. The control parameters ANGLE FEEDBACK, ANGULAR VELOCITY FEEDBACK, POSITION FEEDBACK, and VELOCITY FEEDBACK can be changed for every simulation run.

To be able to examine the response of the system to random perturbations, a *perturbation* is introduced representing a random force acting horizontally on the carriage (as does the *control function*). At each point in time it is determined by random generator. The maximum AMPLITUDE of this *perturbation* is specified by the program user.

For better assessment of the control success the currently required *control power* is computed; it is the product of *control function* and *velocity*. The integral of *control power* over time yields the total *control work* required (equivalent to energy consumption); this is therefore employed as further state variable.

Parameters and initial values
AMPLITUDE = 0 [m*kg/(Second*Second)]
GRAVITATION CONSTANT = 9.81 [m/(Second*Second)]
PENDULUM LENGTH = 1 [m]
PENDULUM MASS = 1 [kg]
RADIUS OF PENDULUM MASS = 0.05 [m]
CARRIAGE MASS = 1 [kg]
INITIAL ANGLE = 0.2 [1]
        other initial values = 0, cf. INTEG equations

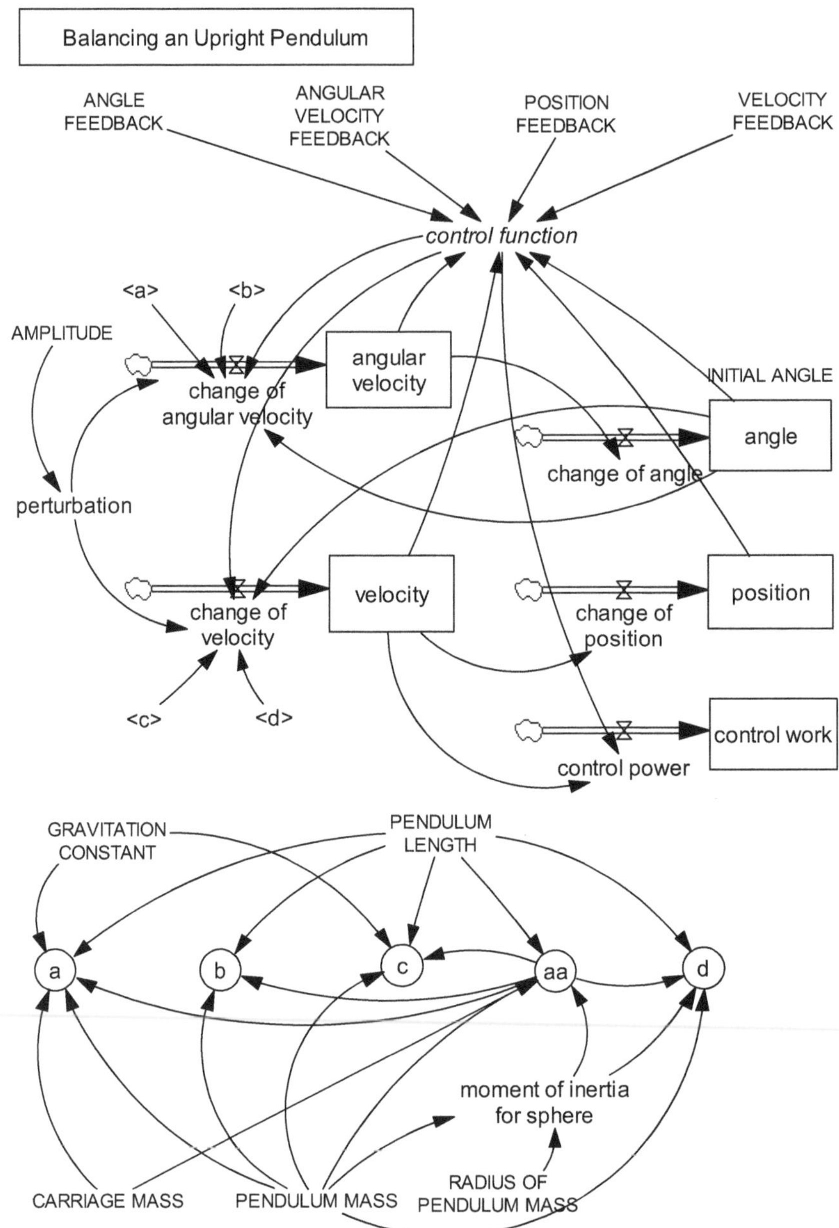

Figure Z209b: Simulation diagram for the upright pendulum with control system.

Computed parameters

moment of inertia for sphere = 2*PENDULUM MASS *RADIUS OF PENDULUM
     MASS^2 /5 [m^2*kg]

aa = moment of inertia for sphere *(PENDULUM MASS +CARRIAGE MASS)
     +PENDULUM MASS *CARRIAGE MASS *PENDULUM LENGTH^2
     [m*m*kg*kg]

a = GRAVITATION CONSTANT *(PENDULUM MASS +CARRIAGE MASS)
     *PENDULUM MASS *PENDULUM LENGTH/aa [1/(Second*Second)]

b = -PENDULUM MASS*PENDULUM LENGTH /aa [1/(m*kg)]

c = -GRAVITATION CONSTANT *PENDULUM MASS^2 *PENDULUM LENGTH^2
     /aa [m/(Second*Second)]

d = (moment of inertia for sphere +PENDULUM MASS *PENDULUM LENGTH^2) /aa
     [1/kg]

Control parameters and control functions

POSITION FEEDBACK = 3 [kg/(Second*Second)]

VELOCITY FEEDBACK = 10 [kg/Second]

ANGLE FEEDBACK = 100 [m*kg/(Second*Second)]

ANGULAR VELOCITY FEEDBACK = 30 [m*kg/Second]

control function = ANGLE FEEDBACK *angle +ANGULAR VELOCITY FEEDBACK
     *angular velocity +POSITION FEEDBACK *Position +VELOCITY FEEDBACK
     *velocity [m*kg/(Second*Second)]

perturbation = AMPLITUDE*(2*RANDOM UNIFORM (0, 1, 0) -1)
     [kg*m/(Second*Second)]

Dynamics

change of velocity = c*angle +d*(control function +perturbation) [m/(Sec^2)]

change of position = velocity [m/Second]

change of angular velocity = a*angle +b*(control function +perturbation) [1/(Sec^2)]

change of angle = angular velocity [1/Second]

velocity = INTEG (change of velocity, 0) [m/Second]

position = INTEG (change of position, 0) [m]

angular velocity = INTEG (change of angular velocity, 0) [1/Second]

angle = INTEG (change of angle, INITIAL ANGLE) [1]

control power = ABS (control function *velocity) [m*m*kg/(Second*Second*Second)]

control work = INTEG (control power, 0) [m*m*kg/(Second*Second)]

Simulation time parameters

INITIAL TIME = 0 [Second]

FINAL TIME = 10 [Second]

SAVEPER = TIME STEP [Second]

TIME STEP = 0.01 [Second]

Simulation results

The default parameter settings of the model correspond to a relatively good solution of the control problem for the upright pendulum. Figure Z209c shows the corresponding dynamic behavior: After an initial perturbation (tilt angle of 0.2 = 11.5 degrees) the falling motion of the pendulum is quickly stopped by the motion of the carriage. After about 2 seconds, the pendulum is stabilized in the vertical position and the carriage is moved back again to the starting point after about 10 seconds.

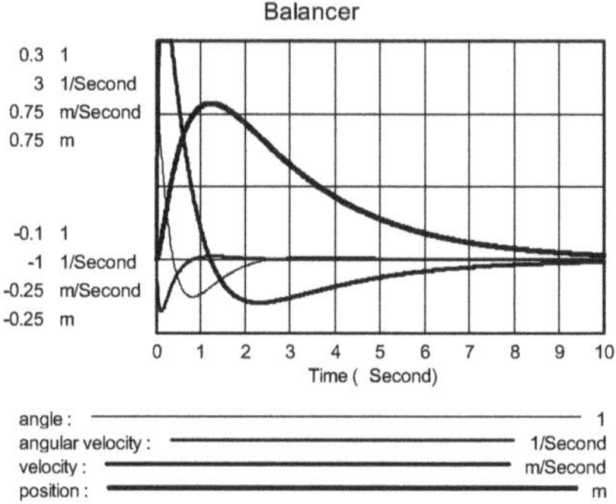

Figure **Z209c**: Stabilization as function of time after strong initial perturbation.

Likewise, the state diagram of *angle* and *position* (Figure Z209d) clearly shows that the falling motion is quickly stopped, and the pendulum is brought back to the upright position (*angle* = 0) and to its initial location (*position* = 0). If a random *perturbation* is introduced in addition (AMPLITUDE = 10; Figure Z209e), the *control function* also manages to bring the pendulum quickly back to the initial position and attitude despite massive perturbations. As the AMPLITUDE of random perturbations increases, effective control becomes more and more difficult, until finally the system no longer succeeds in keeping the pendulum upright.

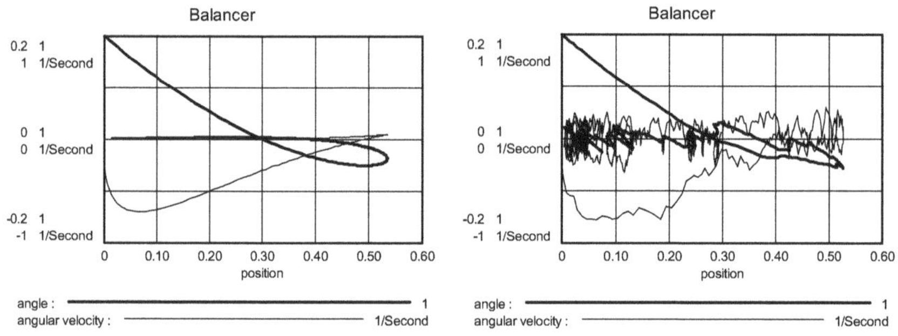

Figure **Z209d**: Stabilization of angle and angular velocity in the state diagram.
Figure **Z209e**: Stabilization of a strong random perturbation (AMPLITUDE = 10).

Further investigations of this system, in particular for finding optimal control parameters, are found in Bossel 1994. It turns out that the choice of control parameters leads to stable solutions only in certain ranges, and that changes of the control parameters even within these limits can result in considerably different dynamic behavior. This becomes quite obvious in experiments with the four control parameters. Change of the pendulum mass also causes a strong change of dynamics of the system.

Exercises

1. Initially, try to control the system entirely by ANGLE FEEDBACK (proportional control: P-control). Change only this control parameter, setting the other three control parameters to 0. Determine the control parameter range for which (a) stability, and (b) instability results. Find the critical parameter value separating the two regions, and determine where the initial tilt angle remains constant (= INITIAL ANGLE). Does the stabilization of the *angle* also result in stabilization of *position*? Discuss what this proportional control can at best achieve with regard to the tilt *angle* of the pendulum.

2. Improve control performance by adding to the proportional control of the *angle* the differential control for the derivative of the angle (i.e. the *angular velocity*) (D-control), using control parameter ANGULAR VELOCITY FEEDBACK. (This combination is referred to as PD-control). The additional use of the *angular velocity* (of tilt) for control now means that the controller can already respond if the tilt is still small, because the *angular velocity* already indicates impending change. Find a combination of the two control parameters which leads to rapidly damped stable behavior. What do you observe with respect to the carriage position? Can you improve position control by a particular combination of both angle parameters?

3. Examine systematically the influence of the two control parameters on frequency and damping of the controlled system with respect to *angle*. Draw the regions of stability in a diagram (ANGULAR VELOCITY FEEDBACK over ANGLE FEEDBACK) and sketch the system response as a function of the two parameters in the diagram.

4. Now use also the other two control parameters for POSITION FEEDBACK and VELOCITY FEEDBACK. Find control parameter combinations which bring the pendulum back into the vertical position and to the starting point of the carriage as quickly and with as much damping as possible.

5. Examine whether the control function found (in the previous exercise) can still bring about system stability at perturbations by a random force with a maximum AMPLITUDE of 10 Newton [1 N = 1 kg m sec^{-2}]. If this is not the case, determine corresponding values for the control parameters. *Note*: Use also the state diagram in assessing the quality of the controller.

6. Replace the linear formulation of the *control function* used so far by a (nonlinear) on-off controller (bang-bang control) for each of the four state variables. This controller does not respond to small excursions from the zero value of the state variable, but reacts with maximum (positive or negative) control force to deviations outside of this zero range. Use the four control parameters to find the maximum control force for acceptable and stable behavior with respect to each of the four state variables (formulate *control function* accordingly).

7. Use other nonlinear expressions for the control function (e.g. quadratic or cubic expressions) and examine the influence on control response. Try to choose control function and control parameters such that the control task is solved with minimal expenditure of energy (i.e. *control work*).

References

Bossel, H. 1992: *Simulation dynamischer Systeme – Grundwissen, Methoden, Programme*. Vieweg Braunschweig und Wiesbaden, 2. Aufl.
Bossel, H. 1994: *Modeling and Simulation*. A K Peters, Wellesley MA.

Bossel, H. 2004: *Systeme, Dynamik, Simulation – Modellbildung, Analyse und Simulation komplexer Systeme*. Books on Demand, Norderstedt.
Bossel, H. 2007: *Systems and Models - Complexity, Dynamics, Evolution, Sustainability*. Books on Demand, Norderstedt.

Z210 Optimizing glider flight: search for thermals

Simulation task

Motorless flight – whether with sailplane, hang-glider or paraglider –challenges the pilot to find invisible updrafts and use them efficiently for gaining altitude and continuing the flight. Regions of best climb must be found quickly, and the pilot should be able to stay in them even if the region of rising air shifts or moves with the wind. Narrow columns of rising warm air called "thermals" are the most common source of lift. Since the pilot cannot see where the air rises, let alone where its rate of ascent is greatest, this task at first seems to be solvable only by luck and chance. Air usually rises fastest in the center of a thermal, and hence the task usually amounts to finding the center and staying in it, i.e. "centering" the thermal.

What would an intelligent blind man do, who is abandoned in a hilly area unknown to him and given the task of scaling the next mountain as quickly as possible? He would probably stop and turn around if he notices that he is going downhill, would search where the path leads uphill, and would continue to climb where the path seems to be the steepest. He will finally recognize the peak by the fact that his steps lead him downhill in all directions. The example shows that it is possible to find an invisible "optimum" by applying a suitable search rule. This optimization task is actually an everyday problem in many applications. If a simulation program is available, mathematical optimization procedures can help in the search for an optimum. These procedures make use of different rules. The rule of "steepest ascent" (used here by the blind man) is one of the most frequently used rules for optimization.

Pilots apply somewhat different rules in searching for and centering thermals (Hesse 1978, Reichmann 1982, Bossel 1998). Whether a rule is better or worse can hardly be determined in practice since conditions in thermals differ. However, search rules can be compared with the help of simulations since conditions in the thermal can then be defined exactly. Using a simulation model we want to examine in the following whether regions of lift can be reliably found and efficiently explored by using a simple flight rule, even if the core of the thermal (with maximum lift) is constantly shifting, and the pilot has no idea where the core is exactly. The mathematical equations for level flight and for turns as well as the decision rules for the thermal search are programmed in the simulation model. The (three-dimensional) spatial distribution of the velocity field of rising air is provided in the computer program. The (three-dimensional) flight paths of the glider resulting from the application of different rules can then be calculated, graphically represented and compared. In particular, it is possible to determine which flight rule produces the greatest height gain in the shortest time – important information for a glider pilot.

Such a simulation model should represent flight maneuvers and flight performance as realistically as possible. The simulation model used here accounts for (among other items): bank angle as a function of flight speed and radius of turn, and resulting increase in flight speed and rate of sink, as well as the time delays of variometer reading and pilot response. (The variometer is an instrument showing current rate of climb or sink of the aircraft.) Normal flight speed, rate of sink and angular velocity of turns (i.e. how many seconds does it take to complete a circle?) are provided as parameters. In this way it is possible to simulate thermal flight of soaring birds, paragliders, hanggliders, and sailplanes. Strong and weak, narrow and wide thermals can be specified by proper choice of parameters specifying location, shape, and strength of the column of rising air as well as its possible inclination from the vertical. Initial position, altitude, and heading can be specified. The (differential) equations of glider flight dynamics are extremely simplified for the purpose of this model (in contrast to the exact equations of motion in model Z211 "Flight dynamics").

Simulation model

The complete simulation program is listed in the following (comments in {brackets}). The corresponding simulation diagram is shown in Figure Z210a. Parameters describing the thermal (RADIUS THERMAL, INCLINATION THERMAL, MAX RATE ASCENT) can be specified in the box "Thermal Parameters". Parameters describing the normal flight conditions of the aircraft (VNORM, VSINK, TIME CIRCLE) are provided in the box "Flight Parameters". Parameters giving the initial position (INITIAL HEADING, BEGIN Y (y is South-North axis) are specified in the box "Position Parameters". (The initial position on the x-axis (West-East axis) is 0; the initial altitude *height* is $z = 1000$ m). Internally, the model uses polar coordinates; the Cartesian coordinates are therefore recalculated. The computation of flight speed and rate of sink must account for the fact that both change as a result of the angle of bank in turns. Whether the aircraft turns or flies level in a thermal depends on the result of the decision rule employed, which accounts for current rate of sink and the rate of change of the rate of sink. Two alternative flight rules are specified below (under the heading "Climbing in a thermal; search rule"). During the simulation, banks and turns are constantly modified according to the flight rule. Heading and position in three-dimensional space are constantly recomputed, and the flight trajectory during the search and centering process is recorded and can be drawn in two- or three-dimensional graphs.

Constants
PI = 3.14159 [1]
DEGREE PER CIRCLE = 360 [degree]
A = degree per circle/(2*pi) [degree]
G = 9.81 [m/(Second*Second)] {*gravitational acceleration*}
RINF = 1e+006 [m]

Thermal parameters
MAX RATE ASCENT = 5 [m/Second]
      {*maximum rate of ascent of air in center of thermal, meters/sec*}
RADIUS THERMAL = 150 [m] {*radius of thermal, meters*}
INCLINATION THERMAL = 45 [degree] {*inclination of thermal from vertical, degrees*}
THETA = inclination thermal/a [1]

Aircraft
VNORM = 10 [m/Second] {*normal flight velocity, meters/sec*}
VSINK = 1 [m/Second]
      {*rate of sink in straight and level flight at normal velocity, meters/sec*}
TIME CIRCLE = 10 [Second] {*time for full circle, seconds*}
TURN = 2*PI /TIME CIRCLE [1/Second]
RETARD = 1 [Second] {*delay*}

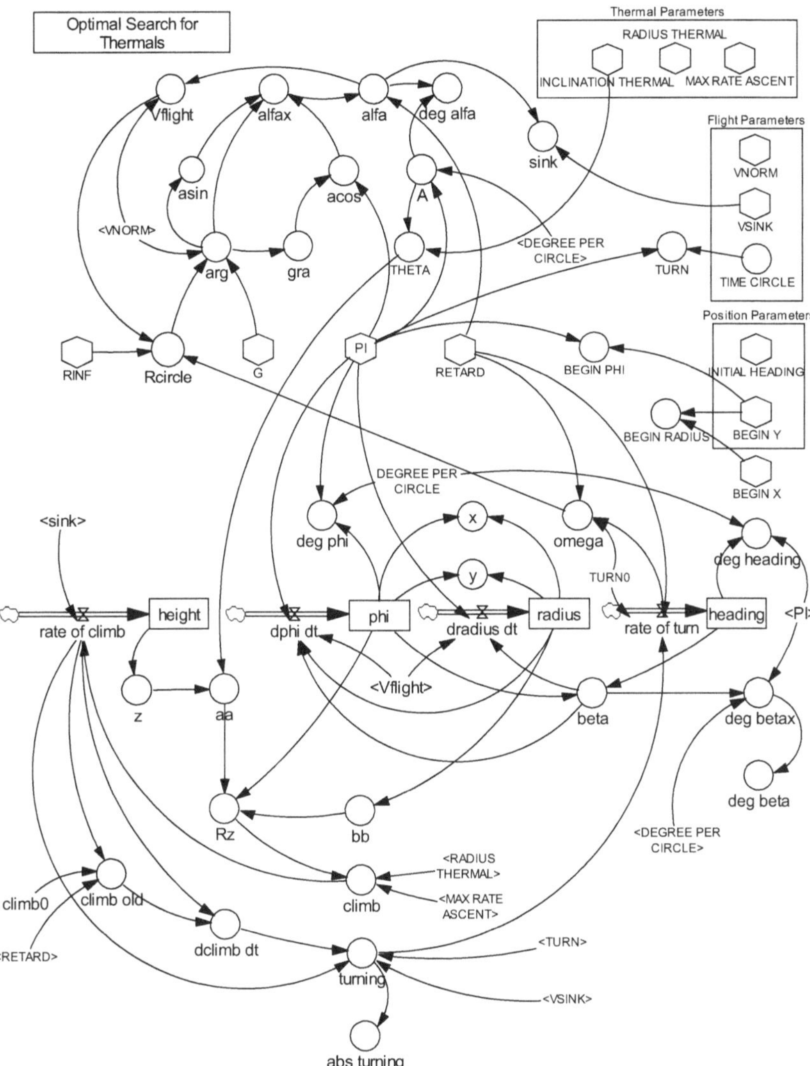

Figure Z210a: Simulation diagram for searching and centering thermals.

Initial position

INITIAL HEADING = 90 [degree]
     {*initial heading: north = 0, east = 90, south = 180, west = 270*}

BEGIN X = 0 [m] {*X initial position, meters east of coordinate origin; do not change!*}

BEGIN Y = 75 [m] {*Y initial position, meters north of origin*}

BEGIN RADIUS = SQRT (BEGIN X *BEGIN X +BEGIN Y *BEGIN Y) [m]

BEGIN PHI = IF THEN ELSE (BEGIN Y>=0, pi/2, -pi/2)

TURN0 = 0 [1/Second]

Flight velocity, roll angle, rate of sink
Rcircle = IF THEN ELSE ((ABS(omega) > 0.001), Vflight /ABS(omega), RINF) [m]
arg = VNORM *VNORM /(Rcircle*G) [1]
gra = 1/arg [1]
asin = arg +arg^3/6 +(3/40)*arg^5 [1]
        {arc sin approximated by 3 terms of Taylor series}
acos = PI/2-(gra +gra^3/6 +(3/40)*gra^5) [1]
        {arc cos approximated by 3 terms of Taylor series}
alfax = IF THEN ELSE ((arg< = 1), asin, acos) [1]
alfa = DELAY1I (alfax, RETARD, 0) [1] *{bank angle}*
deg alfa = alfa*a [degree]
Vflight = VNORM *SQRT (1/COS(alfa)) [m/Second]
sink = VSINK /COS(alfa) [m/Second]

Climbing in thermal; search rule
height = INTEG (rate of climb,1000) [m] *{flight altitude}*
z = height-1000 [m]
aa = 200+(z *(SIN (theta) /COS(theta))) [m] *{thermal begins 200 m east of origin}*
bb = radius [m]
Rz = SQRT (aa*aa +bb*bb -2*aa*bb*COS(phi)) [m]
climb = IF THEN ELSE ((Rz < RADIUS THERMAL), MAX RATE ASCENT*(1 -(Rz /RADIUS
        THERMAL)), 0) [m/Second]
        {strength of thermal increases linearly from outer edge to center, where it is
        maximum}
rate of climb = climb -sink [m/Second]
climb0 = -1 [m/Second]
climb old = DELAY3I (rate, retard, climb0) [m/Second]
dclimb dt = rate of climb –climb old [m/Second]
turning = IF THEN ELSE (((dclimb dt <0) :AND: (rate of climb> -0.5*VSINK)), -TURN, 0)
        [1/Second]
        *{This is **Decision Rule 1** for locating core of thermal: If rate of climb decreases*
        and rate of sink is less than normal sink rate, then 'turn'; else fly straight ahead.
        *Alternative **Decision Rule 2**: Decrease bank angle in turns if thermal lift im-*
        proves; bank more steeply if it deteriorates. Corresponding equations:
        turning1 = IF THEN ELSE (((dclimb dt < 0) :AND: (rate of climb > -0.5*VSINK)), -
        TURN, 0)
        turning = IF THEN ELSE (((dclimb dt >= 0) :AND: (rate of climb > -0.5*VSINK)), -0.3
        TURN, turning1) }
abs turning = ABS (turning) [1/Second]
rate of turn = DELAY1I (turning, RETARD, TURN0) [1/Second]
omega = DELAY1I (rate of turn, RETARD, TURN0) [1/Second]

Heading and position
heading = INTEG (rate of turn, INITIAL HEADING/A) [1] *{heading angle}*
deg heading = (heading/(2*PI) –INTEGER (heading/(2*PI))) *DEGREE PER CIRCLE [degree]
beta = phi +heading [1]
deg betax = (beta/(2*PI) –INTEGER (beta/(2*pi))) *DEGREE PER CIRCLE [degree]
deg beta = IF THEN ELSE ((deg betax > 180), deg betax -360, deg betax) [degree]
phi = INTEG (dphi dt, BEGINPHI) [1] *{location in polar coordinates: angle}*
deg phi = (phi/(2*PI) –INTEGER (phi/(2*PI))) *DEGREE PER CIRCLE [degree]
dphi dt = (Vflight/radius) *SIN((PI/2) -beta) [1/Second]
radius = INTEG (dradius dt, BEGIN RADIUS) [m] *{location in polar coordinates: radius}*
dradius dt = Vflight*COS((PI/2) -beta) [m/Second]

x = radius*cos(phi) [m]
y = radius*sin(phi) [m]

Simulation time parameters
INITIAL TIME = 0 [Second]
FINAL TIME = 250 [Second]
TIME STEP = 0.25 [Second]
SAVEPER = TIME STEP [Second]

Simulation results

In the simulations two alternative flight rules are employed and compared. Rule 2 corresponds to the flight rule favored by former world champion sailplane pilot Reichmann.

Rule 1*: In an area of lift, fly straight ahead if and as long as lift improves or remains constant. Fly a narrow turn if and as long as lift deteriorates.*
Rule 2*: In an area of lift, circle in a flat turn if and as long as lift improves or remains constant. Fly a narrow turn if and as long as lift deteriorates.*

In the simulations a circular area of ascending air is assumed. In the thermal, the rate of ascent rises linearly from the outer edge up to the center (Figure Z210b). The aircraft can climb only in the inner range where the rate of ascent of the rising air is larger than the sinking speed of the glider. To check whether the flight rules can also deal with strongly inclined thermals, the vertical axis of the thermal is tilted by 45 degrees in the default parameter setting of INCLINATION THERMAL.

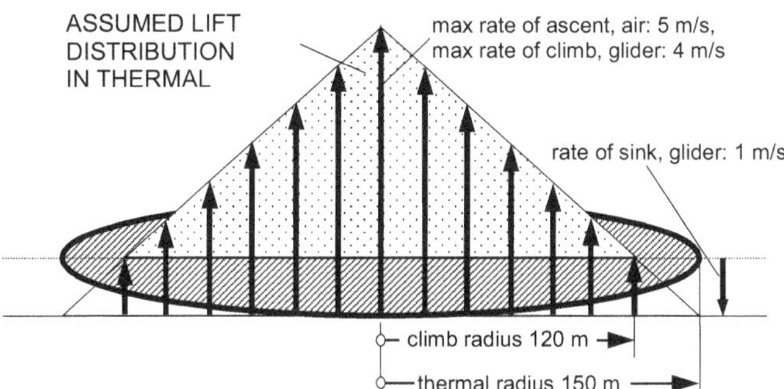

Figure Z210b: Assumed vertical velocity distribution (rate of ascent) in the thermal.

Aircraft can be differentiated primarily by their normal flight speed VNORM: A large bird soars with a speed of about 5 m/s (18 km/h), a paraglider with 7.5 m/s (27 km/h), a hang-glider with 10 m/s (36 km/h) and a sailplane with about 25 m/s (90 km/h). Simplifying somewhat, a rate of sink VSINK of 1 m/s is assumed in the following simulations for these aircraft at their normal flight speed . (The parameter VSINK in the "Flight Parameters" box can be easily changed.) For narrow turns the angular velocity assumed for both flight rules corresponds to an 8 second circle. This value has proved to be optimal in the simulations. It corresponds to a bank angle of 22 degrees for the soaring bird, 32 degrees for the paraglider, 40 degrees for the hang-glider and more than 60 degrees for the sailplane. For the flat turns required in Rule 2, a 30 second circle is optimal, corresponding to a bank angle of 7 degrees for the soaring bird, 9 degrees for the paraglider, 12 degrees for the hang-glider and 30 degrees for the sailplane. To test the performance of the two flight rules for the case where the pilot turns in the "wrong" direction during the search (i.e. away from the thermal), only left turns are permitted in the simulations.

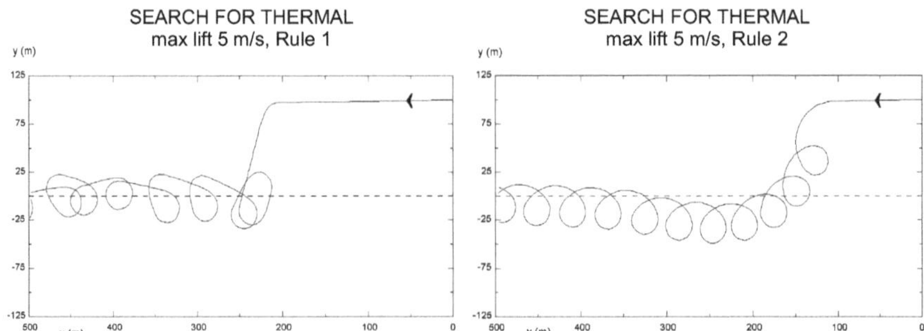

Figure Z210c: Thermal search pattern for Rule 1.
Figure Z210d: Thermal search pattern for Rule 2.

Figures Z210c and d show the flight paths of a hang-glider corresponding to the two flight rules for finding and centering a thermal (having a vertical velocity distribution as in Figure Z210b). In this case the thermal has a RADIUS THERMAL of 150 m with a MAX RATE OF ASCENT of the rising air of 5 m/s. For a rate of sink of 1 m/s the glider therefore finds lift within a radius of 120 m, with a maximum rate of climb of 4 m/s. At an altitude of 1000 m the center of the thermal is located at 200 m in flight direction (with initial heading of 270 degrees, i.e. to the west) and 100 m to the left of the starting point. The thermal has an INCLINATION THERMAL of 45 degrees, tilting west. This means that the centers of the search circles have to be constantly shifted as the glider climbs in the thermal (here: to the west).

Both flight rules are successful; in both cases the thermal is found and centered. But while Rule 1 leads to the center rather quickly and accurately, Rule 2 requires more search circles (meaning loss of time and less gain in altitude). As a result the height reached after 250 seconds is also different: 1663 m with Rule 1, 1609 m with Rule 2. The conditions for flying in the thermal average out to a 14-second circle and 24 degree bank for Rule 1, and a 12-second circle with a (correspondingly steeper) bank of 27 degrees for Rule 2.

What happens if in the initial searching turn the glider flies away from the center of the thermal? (A common occurrence since the pilot cannot see the thermal.) Figures Z210e and f show the search and flight patterns for a glider under the same conditions as before, but with starting positions to the left and right of the center of the thermal. A significant difference between the two flight rules becomes obvious now: Using Rule 2, the center is found only if initial searching turns are in the direction of the center. By contrast, application of Rule 1 will quickly find the center even if the initial searching turn leads away from it. The span of the flight corridor within which the search will lead to the center is significantly wider for Rule 1 than for Rule 2. This means that a pilot using Rule 1 will have a very significant advantage over another pilot applying Rule 2.

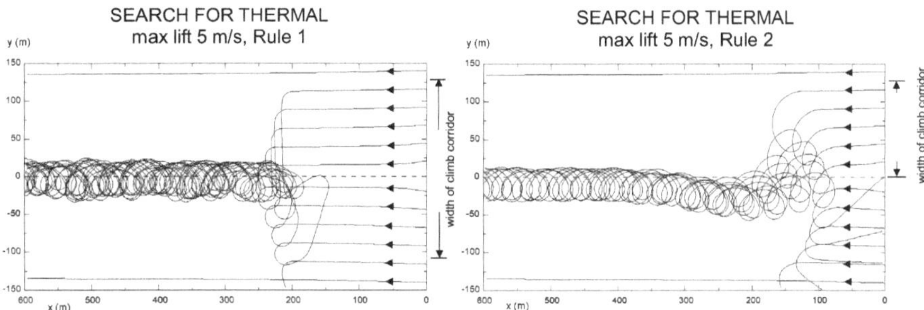

Figure Z210e: The flight corridor within which the center is safely found is almost twice as wide in Rule 1 as it is in Rule 2.
Figure Z210f: Rule 2 does not lead to the thermal if the first searching turn leads away from the center. Overall, the chance for finding the thermal is only about one half of that for Rule 1.

How successful are these search rules for other types of aircraft? Figures Z210g and h show the flight paths for a soaring bird, paraglider, hang-glider and sailplane for the same conditions of thermal lift as before. Again, the differences between the rules are evident: Using Rule 1 the thermal is centered more quickly and more accurately. Within the same time span (250 seconds) a greater height is reached than using Rule 2. The comparatively large turning radius of the sailplane forces it to fly in the outer range of the updraft where the rate of ascent of air is lower. In addition, its sink rate is higher corresponding to its steeper bank angle. Comparison of the flight paths for the soaring bird also shows that Rule 1 probably captures its natural (observed) flight behavior more accurately than Rule 2.

Additional simulations for other lift conditions (narrow and wide, strong and weak, vertical and inclined thermals) produce the same result: Using Rule 1 the thermal is found quickly and centered accurately, even in cases where one would "lose" the thermal if flying according to Rule 2. Most importantly, the flight corridor for successfully locating and using the thermal is (almost) twice as wide for Rule 1 than for Rule 2. In practical terms, Rule 1 translates into the following recommendations for searching for, finding, and centering thermals:

- If rough air, reduced rate of sink, or occasional lift indicate that an area of lift may be close, continue to fly straight as long as lift improves (or the rate of sink decreases) or the variometer reading remains the same.
- As soon as lift decreases (or the rate of sink increases), fly a narrow and steep turn. The direction of the initial turn hardly matters.
- As lift increases again (or the rate of sink decreases) fly straight ahead until lift begins to decrease again (or the rate of sink increases).
- When circling in the thermal, correct the flight path by smoothly reducing or increasing the angle of bank, depending on whether lift increases or decreases.

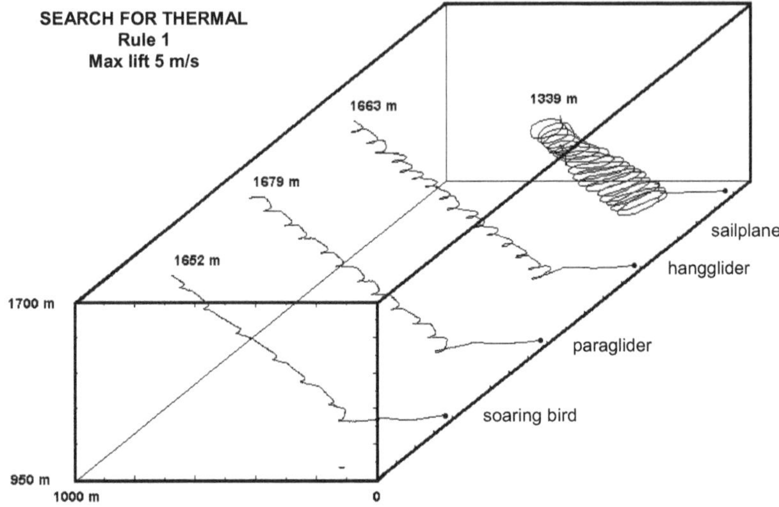

Figure Z210g: Climbing in an inclined thermal according to Rule 1.

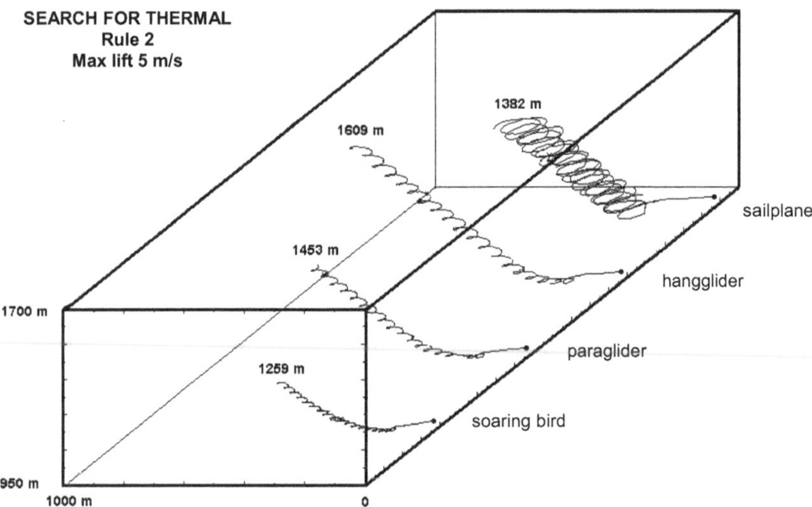

Figure Z210g: Climbing in an inclined thermal according to Rule 2.

The essential rule in an area of lift is therefore:

- ***Bank and turn (steeply) as soon as the variometer reading deteriorates; fly straight ahead as long as it improves or stays constant.***

It is therefore primarily important to pay attention to the *change* of the variometer reading, and less so to the rate of climb or sink itself.

Exercises

1. Choose parameters (VNORM, VSINK, TIME CIRCLE) for a particular aircraft type (see text) and a vertical thermal (INCLINATION THERMAL = 0). By changing the initial position, investigate in what region the thermal can still be safely found and efficiently used, initially using Rule 1, then Rule 2. Discuss the result and draw conclusions as recommendations to pilots.

2. Compare flight paths and maximum heights reached after 250 seconds for (a) different aircraft types, and (b) different angles of INCLINATION THERMAL. What general conclusions can be drawn from this?

3. Investigate if and under which conditions for the different thermal parameters (MAX RATE ASCENT, RADIUS THERMAL, INCLINATION THERMAL) for weak, strong, narrow, wide, vertical, and inclined thermals Rule 1 or Rule 2 provides better performance.

4. Simulate the search and centering process for a thermal with a rectangular lift distribution (constant rate of ascent inside a circular cylinder). Compare results with those found in Exercise 1.

5. Insert an IF-condition in the simulation statements to switch easily to another flight rule before a simulation.

6. Copy the simulation results for *time*, *x*, *y*, and *z* in a three-dimensional graphics program (perhaps written by you), and draw the flight path in three-dimensional space.

7. To simulate hill soaring (i.e. soaring in the rising air upwind of hills), change the area of lift to a rectangular field with a constant or "roof-like" lift distribution. What flight path pattern is generated by applying Rules 1 and 2?

8. To simulate a turbulent thermal, add random bursts of ascending and descending air with adjustable TURBULENCE STRENGTH to the otherwise steady field of rising air. How do the search rules perform under these circumstances? (Use the pulse and random functions of the simulation software to modify the model equations.)

References

Hesse, F., Hesse, W. 1978: *Der Segelflugzeugführer*. Verlag Hesse, Breidenbach.
Reichmann, H. 1982: *Streckensegelflug*. Motorbuch Verlag, Stuttgart, 5th ed.
Bossel, H. 1998: Der beste Weg zum Aufwind. *Fly and Glide*, 5 (71-74), 10 (80).

Z211 Flight dynamics

Simulation task

The simulation of aircraft dynamics using flight simulators is of enormous importance for the development of new aircraft and for pilot training. Before even a prototype is built, a new aircraft design can be thoroughly tested by "flying" it in computer simulations, and using the results to modify the design until it meets the specifications. Pilots can learn to handle all conceivable dangerous situations and maneuvers without endangering themselves and others. The exact simulation of flight dynamics is possible because the mechanics and aerodynamics of the system can be represented by a complex system of differential equations in which a variety of parameters characterize the properties, peculiarities, and performance of the airplane and its aerodynamics.

In addition to aerodynamic efficiency of a high performance sailplane, its response to control inputs by the pilot and its maneuverability are of particular interest. The glider should be able to execute quickly and efficiently and with a minimum loss of altitude the many turns constantly required in soaring flight. Computer simulations of flight dynamics allow: (1) examining aircraft behavior and performance in rapid turn and bank maneuvers; (2) comparing the corresponding behavior of different glider types; and (3) determining the role and influence of the different aerodynamic coefficients ("derivatives") on flight dynamics and performance.

The procedure for setting up the system of differential equations of aircraft motion is briefly described in the following, and is then applied to develop the simulation model for a particular sailplane. The simulation model is used to simulate rapid changes from steeply circling flight in one direction to turning in the opposite direction. The complex system of equations was derived elsewhere (Bossel 1961) using previous work in flight dynamics (Etkins 1956). The flight dynamic derivatives used in the equations were derived (by Heil 1961) for the K6 CR sailplane (designer: Rudolf Kaiser). The procedure used is identical to that applied to any other aircraft, except that the terms relating to engine and propulsion are missing. The simulation model uses the names of variables and parameters as documented in the original publications mentioned.

Deriving the system equations

Moving in three-dimensional space the aircraft (assumed as a rigid body) has nine degrees of freedom (three for translation, three for rotation, three for control surfaces). Its motion is therefore completely described by a system of nine differential equations with nine dependent variables and time t as independent variable. The three types of control surfaces (ailerons, elevator, rudder) control roll motion (about the x-axis, see Fig. Z211a), pitch (about the y-axis), and yaw (about the z-axis). The translational and rotational motions are not independent of each other; yawing motion produces a rolling moment, for example. In the equations of motion, these dependencies are captured by partial derivatives specifying how much rolling moment is caused per unit of yawing motion, for example. Multiplied by current rate of yaw, the respective contribution to roll is computed. Since in flight dynamics the motion with respect to one degree of freedom usually also influences motion with respect to every other degree of freedom, the equations of motion become quite complex and require the specification of a large number of "derivatives", which must be mostly empirically determined. If control inputs are specified, or if they are determined as pilot reaction to aircraft motion, six differential equations remain. These are nonlinear and must be solved numerically, i.e. by simulation.

The equations are developed in an aircraft-fixed orthogonal coordinate system (x, y, z), whose origin is the center of gravity of the airplane (Figure Z211a). The x-axis (longitudinal axis) points to the front, the y-axis (lateral axis) to the right hand wing tip, the z-axis (vertical axis) points downward. The three velocity components of translation (v_x, v_y, v_z) correspond to the directions of these axes. The three angular velocity components around the three axes $(\omega_x, \omega_y, \omega_z)$ are counted clockwise positive. Orientation in space is specified by reference to an earth-fixed coordinate system by pitch angle θ, bank angle φ and yaw angle ψ. The equations of motion are written for a reference state (straight and level flight at normal flight speed). Perturbations from the reference velocity and the reference flow direction are expressed by velocity perturbation u, angle-of-attack perturbation α and yaw perturbation β. The deflection angles of control surfaces are expressed by aileron deflection ξ (causing rotation around x-axis), elevator deflection η (causing rotation around y-axis) and rudder deflection ζ (causing rotation around z-axis).

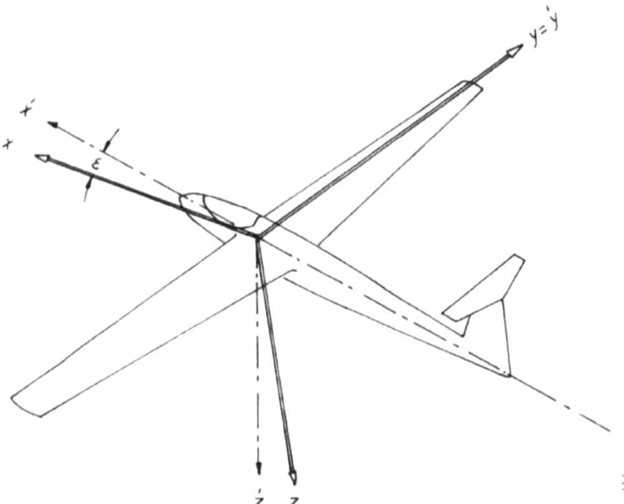

Figure Z211a: Aircraft-fixed orthogonal coordinate system.

The three equations of translational motion with respect to each axis (x, y, z) have the general form:

aerodynamic forces + weight forces = mass forces

and the three equations of rotational motion around the three axes are of the form:

aerodynamic moments = mass moments of inertia

The components of these equations of motion consist of complex expressions in which in particular velocities, accelerations, angular velocities, and angular accelerations of the aircraft must be correctly described for all three axes. For further analysis, dimensionless quantities are introduced. For practical applications the results are later converted back to real quantities.

After lengthy derivation one finally obtains six coupled nonlinear differential equations for the rates of change of the six state variables u, α, β, θ', φ', ψ'. Numerical integration of these rates of change yields the state variables. Further integration with respect to time produces the time behavior for θ, φ, and ψ.

After introducing Taylor series for the variables and linearization, it turns out that the three equations for translational motion (for u, α, β) and the three for rotational motion (for θ', φ', ψ') become decoupled, and can be treated separately in the neighborhood of the reference state. Using Cramer's rule the rates of change of the state variables can be separately determined and numerically integrated for each of these two sets of equations.

For the computation of flight dynamics (particularly the dynamics of rapid turn-and-bank maneuvers) the aileron deflection as function of time is provided as control scenario. This control input causes perturbations of angle of attack α, pitch angle θ (orientation of the pilot with respect to the horizon), and yaw angle β which the pilot has to bring back to zero by appropriate control of elevator and rudder. Appropriate control parameters are provided in the three equations for the control surfaces.

Simulation model

Using current state values, parameters of the aircraft, and aerodynamic coefficients (see below) the right hand sides of the (2 times 3) simultaneous equations for the rates of change are first determined. All dimensions are in mks-units (meters, kilograms, seconds as basic units).

Right hand side of the equations for the rates of change
R1 = (1/EMU) *(-CW +(D7*OX) *OX +(D8 *XI +D9) *XI+(D10 *ZT +D11) *ZT)
    +(1/EMU) *((D1 *AL1DEL +D2 *XI +D3) *AL +(D4 *BT +D5 *BT1DEL +D6 *ZT)
    *BT-CAG *TT) +AL*(TT1 *COFI +PS1 *SIFI) +BT *(PS1 *COFI -TT1 *SIFI)
R2 = (1/EMU) *(D13 *OX +D14 *OZ +D15 *XI +D16 *ZT +D12 *BT +CAG *SIFI) +TT1
    *SIFIPS1 *COFI +AL*(FI1 -PS1 *TT)
R3 = (1/EMU) *(-CA +(D18 *OX) *OX +D19 *XI +D17 *AL +CAG *COFI) +TT1 *COFI
    +PS1 *SIFI +BT*(FI1 -PS1 *TT)
R4 = D24 *OX +D25 *OZ +D26 *XI +(D20 *BT +D21 *BT1DEL +D22 *OZ) *AL +D23
    *BT +T1 *PS1 *TT1 -T6*(TT1 *PS1 *COFI^2 +PS1^2 *SIFI *COFI-TT1^2 *SIFI
    *COFI -TT1 *PS1 *SIFI^2)
R5 = (D28 *OX) *OX +D29 *OY +D30 *ET +D27 *AL +T2*(TT1 *FI1 *SIFI-PS1 *FI1
    *COFI +TT1 *TT *SIFI) -T4*(FI1 *PS1 *COFI -TT1 *FI1 *SIFI -PS1^2 *TT *COFI
    +TT1 *PS1 *TT *SIFI)
R6 = D33 *OX +D34 *OZ +(D35 *XI +D36) *XI +D37 *ZT +(D31 *XI) *AL +D32 *BT
    +T3*(PS1 *FI1 *SIFI +PS1 *TT1 *TT *COFI +FI1 *TT1 *COFI) -T5*(FI1 *TT1
    *COFI -PS1 *TT1 *TT *COFI +FI1 *PS1 *SIFI -PS1^2 *TT *SIFI)

With these results the denominator determinants E4 and E8 and the numerator determinants E1, E2, E3, and E5, E6, E7 for the two sets of equations are determined, and used to compute the rates of change of the state variables using Cramer's rule.

Determinants
E1 = R1 -R2*(AL^2 *BT -BT) +R3 *AL
E2 = -R1*(AL *BT^2 +AL) -2 *R2 *AL *BT +R3*(1 -BT^2)
E3 = -R1 *BT -R2*(1 +AL^2) -R3 *AL *BT
E4 = 1 +AL^2 *BT^2 +AL^2 -BT^2
E5 = -R5 *T1 *T3 *COFI +R6 *T1 *T2 *SIFI
E6 = -T2 *COFI*(R4 *T3 *COFI +T1 *TT *R6) -T3 *SIFI*(R4 *T2 *SIFI +T1 *TT *R5)
E7 = -T2 *COFI *T1 *R6 -T3 *SIFI *T1 *R5
E8 = -T2 *COFI *T1 *T3 *COFI -T3 *SIFI *T1 *T2 *SIFI

Solution of the system of equations
u' = U1 = E1/E4
α' = AL1 = E2/E4
β' = BT1 = E3/E4
θ" = TT2 = E5/E8
φ" = FI2 = E6/E8
ψ" = PS2 = E7/E8

The first three rates of change are integrated once, the last three twice with respect to time to obtain the corresponding (dimensionless) state variables. The index ...X designates the state variables converted to real (dimensional) quantities.

Integration
α = AL = INTEG (AL1, 0)
AL1DEL = DELAY FIXED (AL1, TIME STEP, 0)
ALX = AL *57.3 [degree]

β = BT = INTEG (BT1, 0)
BT1DEL = DELAY FIXED (BT1, TIME STEP, 0)
BTX = BT *57.3 [degree]

θ = TT = INTEG (TT11, -0.0351)
TTX = TT *57.3 [degree]

φ = FI = INTEG (FI11, 0)
FIX = FI *57.3 [degree]

ψ = PS = INTEG (PS11, 0)
PSX = PS *57.3 [degree]

θ' = TT1 = INTEG (TT2, 0)
TT11 = TT1
TT1X = 57.3 *TT1 *V/EL [degree/Second]

φ' = FI1 = INTEG (FI2, 0)
FI11 = FI1
FI1X = 57.3 *FI1 *V/EL [degree/Second]

ψ' = PS1 = INTEG (PS2, 0)
PS11 = PS1
PS1X = 57.3 *PS1 *V/EL [degree/Second]

Aileron deflection as function of time is provided as a scenario; time-dependent elevator and rudder deflections follow from the corresponding control equations and control parameters.

Aileron (scenario)
ξ = XI = DELAY FIXED (AXI *XISCEN, TIME STEP, 0)
XISCEN = WITH LOOKUP (TN,([(0, -2) -(200, 2)], (0, 0), (0, 0), (1, 1), (2, 1), (3, 0),
          (60, 0), (61, -2), (62, -2), (63, 0), (70, 0), (71, -1), (72, -1), (73, 0), (100, 0), (200,
          0)))
AXI = 0.183 [radian] {*aileron angle*}
XIX = XI *57.3 [degree]

Elevator
η = ET = C1 *(AL1DEL +TT1) +C2*(AL +TT)
ETX = ET *57.3 [degree]

Rudder
ς = ZT = C4 *BT1DEL +C5 *BT
ZTX = ZT *57.3 [degree]

Control parameters for elevator, rudder, and ailerons (with reference values Ref)
KDAL = 20 [Ref: 20]
C1 = KDAL
KDBT = 20 [Ref: 20]
C4 = KDBT
KPAL = 0.5 [Ref: 0.5]
C2 = KPAL
KPBT = 0.5 [Ref: 0.5]
C5 = KPBT
KPXI = 0.233

Position, velocity and angular velocities of the aircraft are calculated from the state variables. Natural time is computed from the dimensionless simulation time.

Computation of location
DX = V *COS (PS -BT) *EL/V
X = INTEG (DX, 0)
DY = V *SIN (PS -BT) *EL/V
Y = INTEG (DY, 0)

Flight velocity
V = VDEL *(1 +U1 *TIME STEP)
V0 = 23.6
VDEL = DELAY FIXED (V, TIME STEP, V0)

Angular velocities
ω_x = OX = IF THEN ELSE (Time <= INITIAL TIME, AOX, FI1 -PS1 *TT)
OXX = 57.3*OX*V/EL [degree/Second]
ω_y = OY = IF THEN ELSE (Time <= INITIAL TIME, AOY, PS1 *SIFI)
OYX = 57.3*OY*V/EL [degree/Second]
ω_z = OZ = IF THEN ELSE (Time <= INITIAL TIME, AOZ, PS1 *COFI)
OZX = 57.3 *OZ *V/EL [degree/Second]

Natural time
TN = TNDEL +TIME STEP *EL/V [Second]
TNDEL = DELAY FIXED (TN, TIME STEP, 0)

Initial values, aircraft parameters, and aerodynamic coefficients (Heil 1961) used in the above calculations apply to the K6 CR glider (mass 300 kg). The reference setting of parameters corresponds to flight altitude = 1000 m at flight speed V_0 of 23.6 m/s = 85 km/h, lift coefficient c_a = 0,767 and drag coefficient c_w = 0.0269.

Lift and drag coefficients
ACAG = 0.7676
CA = 0.767
CAG = DELAY FIXED (ROMF/ V^2, TIME STEP, ACAG)
CW = 0.0269

Mass inertias
T1 = 300
T2 = 73.5
T3 = 357
T4 = T1 -T3 = -57
T5 = T2 -T1 = -226.5
T6 = T3 -T2 = 283.5

Parameters
EL = 0.892
EMU = 48.6
ROMF = 426

Aerodynamic coefficients

D01 = -0.224	D11 = 0	D21 = -1.32	D31 = -8.15
D02 = -0.45	D12 = 0.38	D22 = 201	D32 = -0.915
D03 = 0.563	D13 = -1.65	D23 = 2.32	D33 = -8.66
D04 = 0.115	D14 = 1.72	D24 = -108	D34 = -6.12
D05 = -1.57	D15 = 0.058	D25 = 22.5	D35 = 0.63
D06 = -0.12	D16 = 0.222	D26 = 5.46	D36 = -0.975
D07 = 94.25	D17 = -6.128	D27 = -0.926	D37 = -1.01
D08 = -0.325	D18 = -123	D28 = -33.15	
D09 = 0.054	D19 = 0.3	D29 = -8.08	
D10 = -0.06	D20 = -10.1	D30 = -1.28	

Initial values
AOX = 0
AOY = 0
AOZ = 0
ASIFI = 0
SIFI = IF THEN ELSE (Time <= INITIAL TIME, ASIFI, SIN(FI))
ACOFI = 1
COFI = IF THEN ELSE(Time <= INITIAL TIME, ACOFI, COS(FI))
AU = 0

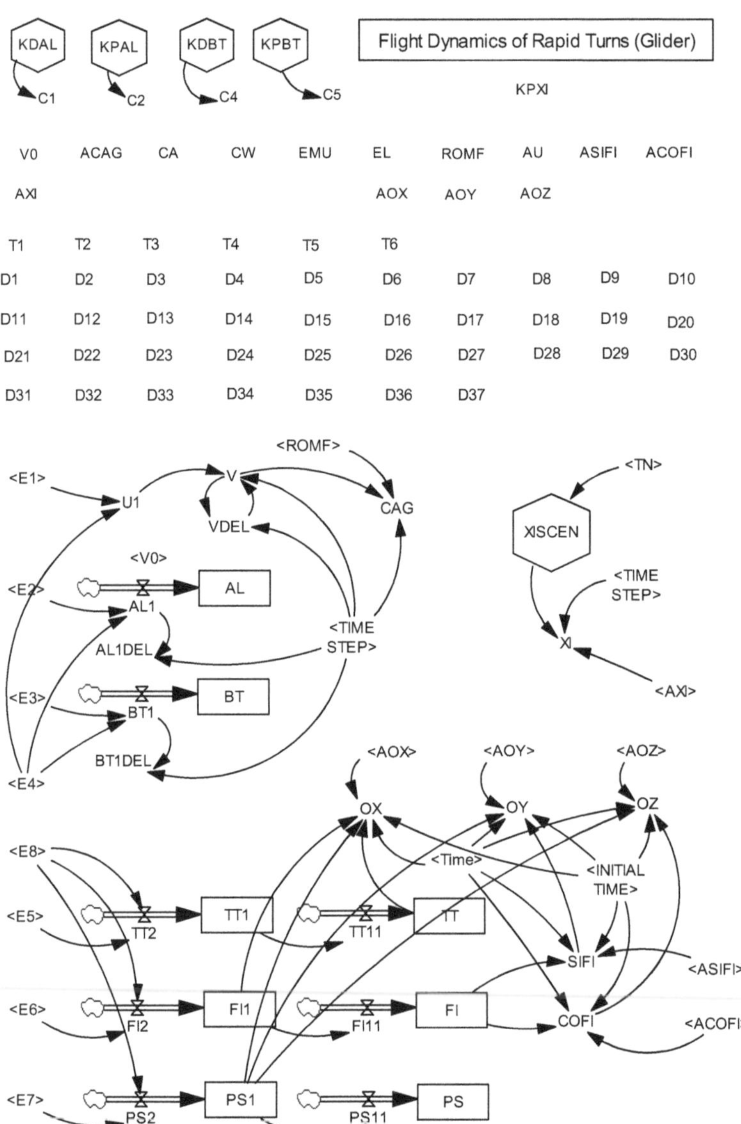

Figure Z211b: Simulation diagram for flight dynamics, part 1.

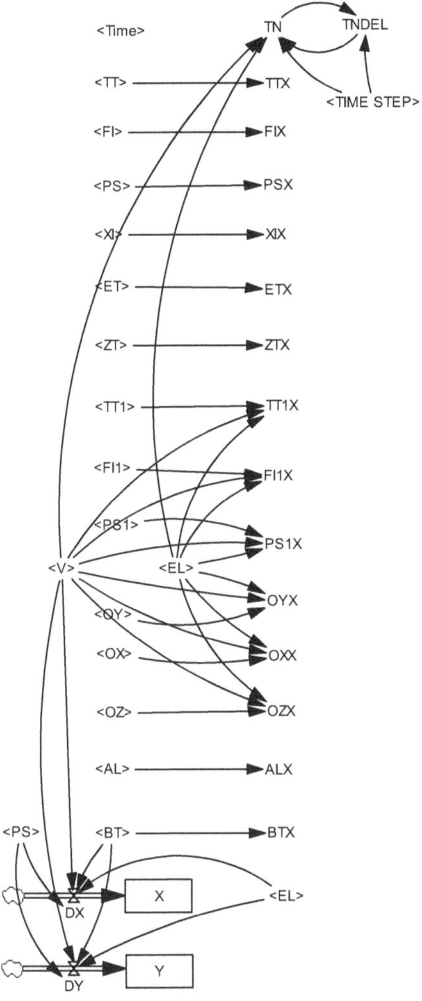

Figure Z211c: Simulation diagram for flight dynamics, part 2.

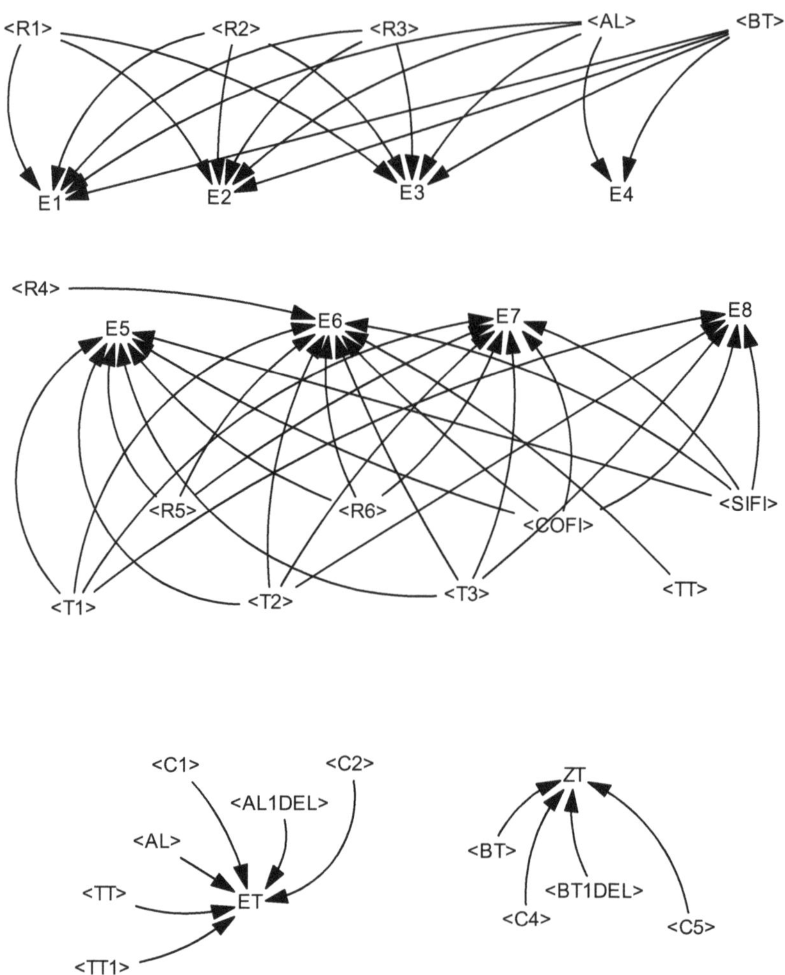

Figure Z211d: Simulation diagram for flight dynamics, part 3.

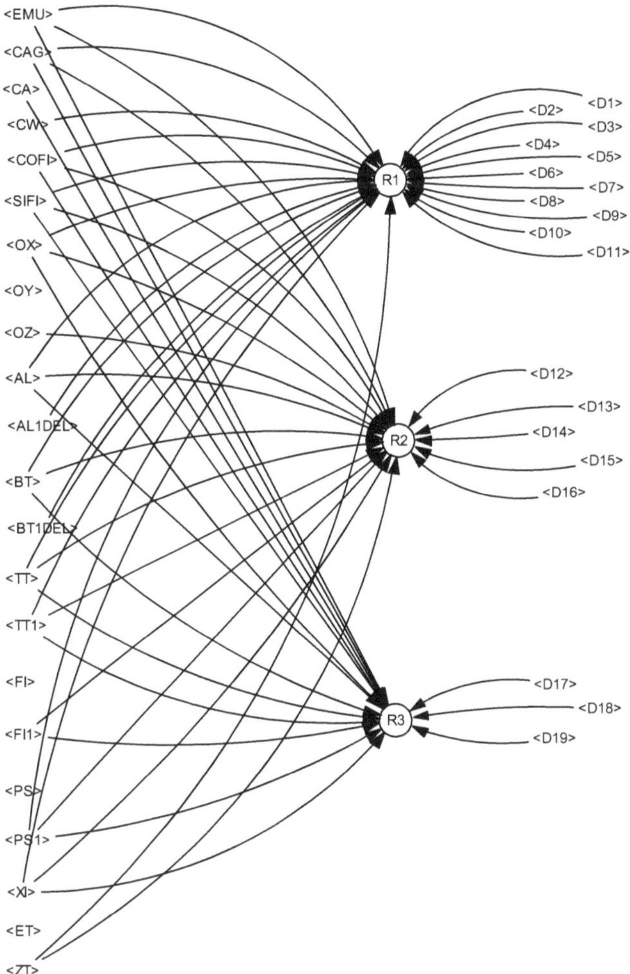

Figure Z211e: Simulation diagram for flight dynamics, part 4.

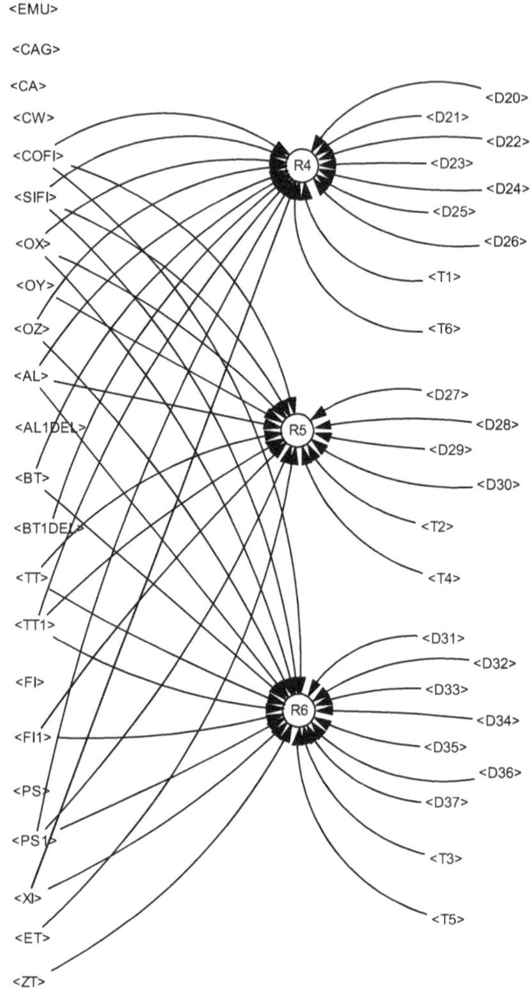

Figure Z211f: Simulation diagram for flight dynamics, part 5.

The system is now specified completely and can be simulated. The corresponding simulation diagrams are shown in Figures Z211b through f. The following time parameters are used for the standard simulation using Runge-Kutta integration.

Simulation time parameters
FINAL TIME = 500 [dimensionless time]
INITIAL TIME = 0 [dimensionless time]
TIME STEP = 0.1 [dimensionless time]
SAVEPER = TIME STEP [dimensionless time]

Simulation results

Figure Z211g shows simulation results for rapid turning maneuvers at different flight speeds (i.e. lift coefficients ACAG). The corresponding control deflections for the reference case (default setting; normal circle) are shown in Figure Z211h and i as function of time. The bank angle during the maneuver is plotted in Figure Z211j.

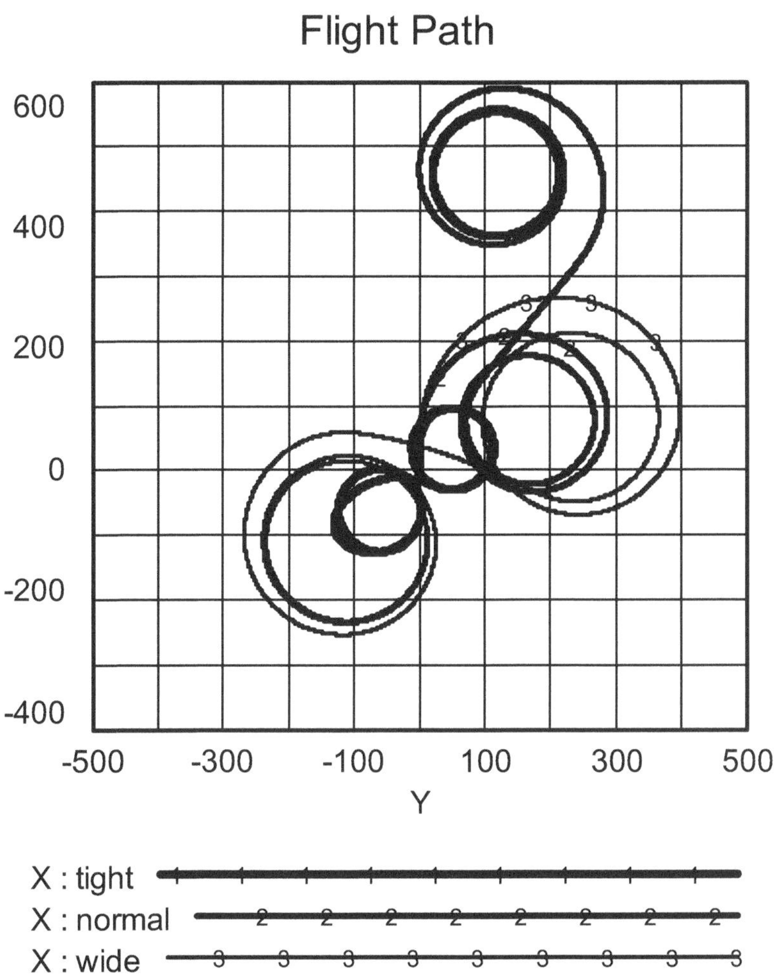

Figure Z211g: Simulation of rapid turning maneuvers for three different cases.
narrow: (narrow circle)   AXI = 0.3,   CA = 1.5
normal: (normal circle)   AXI = 0.183, CA = 0.7676
wide: (wide circle)   AXI = 0.183, CA = 0.5

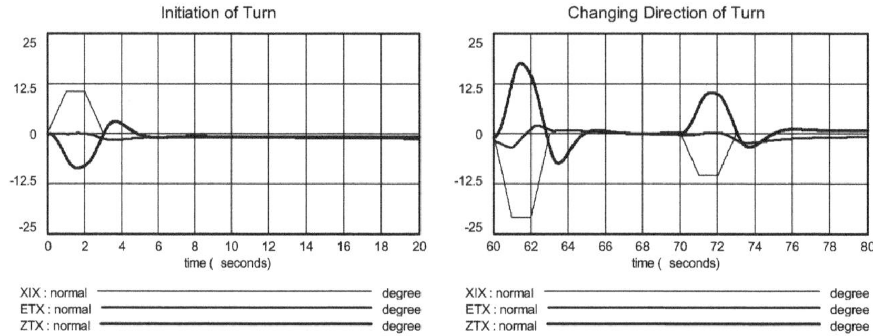

Figure Z211h: Time plot of control adjustment for initiating turn (case: normal circle)
Figure Z211i: Time plot of control adjustment for terminating turn, straightening out, and initiating turn in the opposite direction (aileron XIX, elevator ETX, rudder ZTX).

High lift coefficient means slower flight speed and therefore allows tighter turns. In all three cases in Figure Z211g the initial position is at the coordinate origin, with the glider heading "north". The control scenario is identical for all three cases: A right turn is initiated by applying full (right) aileron for 3 seconds. After 60 seconds the right turn is stopped by full (left) aileron. After a few seconds of straight and level flight, a left turn is initiated (in the same way as the previous right turn) and continued till the end of the simulation (250 seconds).

In all simulations the aircraft quickly and smoothly turns to stable circling flight, and continues to circle with the controls in (almost) neutral setting. The simulation results are confirmed by practical flight experience with the K6 CR sailplane, which for decades was very popular among glider pilots because of its well-tuned controls and excellent performance.

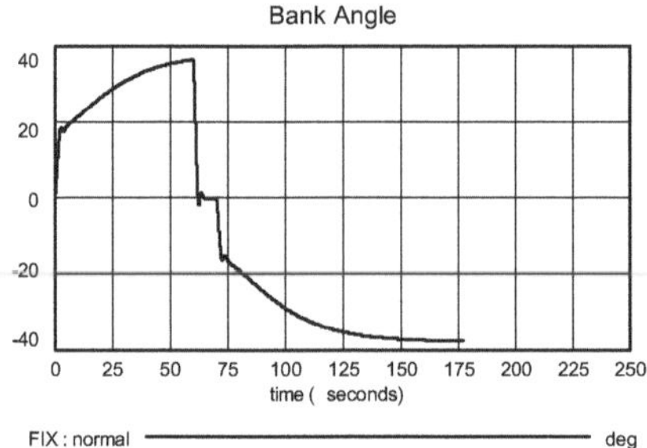

Figure Z211j: Bank angle during the rapid turn maneuver (case: normal circle).

Exercises

1. Examine aircraft behavior in rapid turning maneuvers at different flight speeds (change the lift coefficient ACAG) and for different control scenarios (aileron scenario XISCEN). *Note*: Choose natural time TN as time axis in time plots!

2. Generate time plots of control deflections for initiating turns and for changing the direction of turns. Investigate the effect of different control parameters (KDAL, KPAL, KDBT, KPBT) on "smoothness" and stability of the control process.

3. Produce simulation runs for the three cases shown in Figure Z211g (change of maximum aileron AXI and lift coefficient CA). Generate time plots with three curves each for the control deflections (ailerons *XIX*, rudder *ZTX*, elevator *ETX*), for angular velocities (*OXX, OYX, OZX*), for speed *V*, and for angle of bank *FIX*. Discuss and compare the results.

4. Supplement the simulation program by adding statements for the calculation of the rate of sink and resulting altitude as function of time. Plot the result. Find out how maneuvers for changing the direction of turn can be completed in the shortest time with minimum height loss. (This task requires some knowledge of flight mechanics.)

References

Etkin, B. 1959: *Dynamics of Flight, Stability and Control*. John Wiley, New York.
Bossel, H. 1961: *Berechnung des Kurvenwechselvorgangs von Segelflugzeugen*. Lehrstuhl und Institut für Luftfahrttechnik, Technische Hochschule Darmstadt; Forschungsberichte der Flugwissenschaftlichen Forschungsstelle München der Deutschen Forschungs- und Versuchsanstalt für Luft- und Raumfahrt DFVLR Nr. 102, München.
Heil, W. 1961: *Ermittlung der Derivativa für das Segelflugzeug K 6 CR und Untersuchung einiger Flugzustände*. Lehrstuhl und Institut für Luftfahrttechnik, Technische Hochschule Darmstadt.

Z212 House heating dynamics

Simulation task

A house offers a favorable environment independent of the weather. In cooler climates this means in particular that it must provide a relatively warm environment during the cold season. This is usually possible only by use of energy, i.e. heating. The house and its heating system represent a system providing its residents with the energy service "warm room". This energy service "warm room" can e.g. be described by a measure like "50 cubic meters of space at 20 degrees centigrade for every resident". The example shows that energy services are often quantities which cannot be measured in energy units. The statement of desired energy service leaves completely unspecified how it shall be provided. This may be achieved with an open fire and open window at enormous energy consumption; however, it could also be achieved by clever use of the limited solar energy even in winter in a well insulated house. The purpose of the house and its heating system is therefore in particular to reduce the heat loss to the colder environment and to replace the unavoidable energy loss by heat from an efficient heating source.

If the temperature in the house remains the same, e.g. at a desired room temperature of 20 degrees centigrade, then the energy loss through the outer shell of the house is exactly in flow equilibrium with the energy supplied by the heating system. By balancing all inputs and outflows of energy (except from the heating system) it is therefore possible to calculate the heating energy required for the desired room temperature. The heat losses of a house arise primarily from heat transfer through walls, windows, ceiling and roof as well as floor and basement. These heat transfers can be usefully expressed in watts per square meter of wall area and degree Kelvin temperature difference between inner and outer wall. The corresponding numerical value is designated as k-value. This k-value is small for well insulated walls and large for walls having poor thermal insulation. Further heat losses are associated with the necessary air exchange, by opening outer doors etc. Except from its heating system, however, the house also gains heat from other sources which can play a considerable role in the heat balance particularly for well-insulated houses. Part of this is even the heat output of each resident, amounting to about 100 watts per person. Other sources of heat are the heat losses of lights and appliances as well as the often considerable heat gains by south-facing windows (in the northern hemisphere) as well as solar irradiation on the outer walls.

The balance of heat gains and losses often shows that in particular in early spring and late fall heating is not required although the outside temperature has dropped below the desired room temperature. For well-insulated houses in particular heat gains then outweigh heat losses, and heating becomes necessary only at lower outside temperatures. The outside temperature at which heating becomes necessary is referred to as heating limit temperature. The better the insulation, the lower the heating limit temperature. If the outside temperature falls below this value, then heating is required, where the corresponding energy flow depends of course on the quality of the overall insulation. For the design of a new house or retrofitting of an older building it is of particular interest to determine what energy savings can be achieved by using different insulation measures, and by particular choices of window size and placement (south-facing windows!).

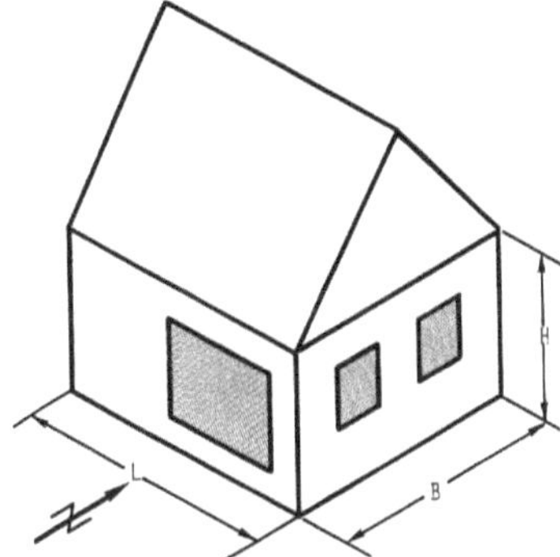

Figure 212 a: House geometry and orientation.

Simulation model

Model Z212 "House heating dynamics" is the (simple) prototype of an interactive simulation model for energy consulting. The model corresponds to a modest rectangular detached house oriented in north-south or east west direction (Figure Z212a). The insulated part of the building is assumed as a cube of length of L, width B and height H. For its six outer surfaces (four side walls, ground floor without basement, ceiling) the window fractions of the walls, and the insulation parameters for windows, walls, floor, and ceiling are specified. For south-facing windows the heat gain is computed as function of seasonally changing solar position. The annual variation of outside temperature, the (constant) temperature of the ground, and the desired room temperature are specified. Heat losses by leaks and airings are not considered in this simple model.

The simulation diagram of the model is shown in Figure Z212b. The complete set of model equations is listed in the following.

The dimensions of the house and the window fractions of walls must be specified first. From this the ceiling area, the floor area, the wall area (without windows), the window area facing south, and the remaining window area can be calculated. The insulation parameters of walls, ceiling, floor, and windows as well as desired room temperature have to be provided.

With these data the heat losses per degree temperature difference can now be computed for ceiling, walls, and windows. From the heat loss to the ground, the heat gain by south windows, and the desired room temperature the heating limit temperature can be determined. It specifies the outside temperature below which heating is required. From the temperature difference to the outside temperature and the furnace efficiency follow the heating power demand, the furnace power demand, and finally the fuel burn rate. If this fuel consumption is integrated over the heating season, one obtains the complete heating fuel consumption. The table functions for solar irradiation through vertical south-facing windows and for outside temperature correspond to a geographical latitude of 50 degrees in Central Europe (but can be easily changed for other regions).

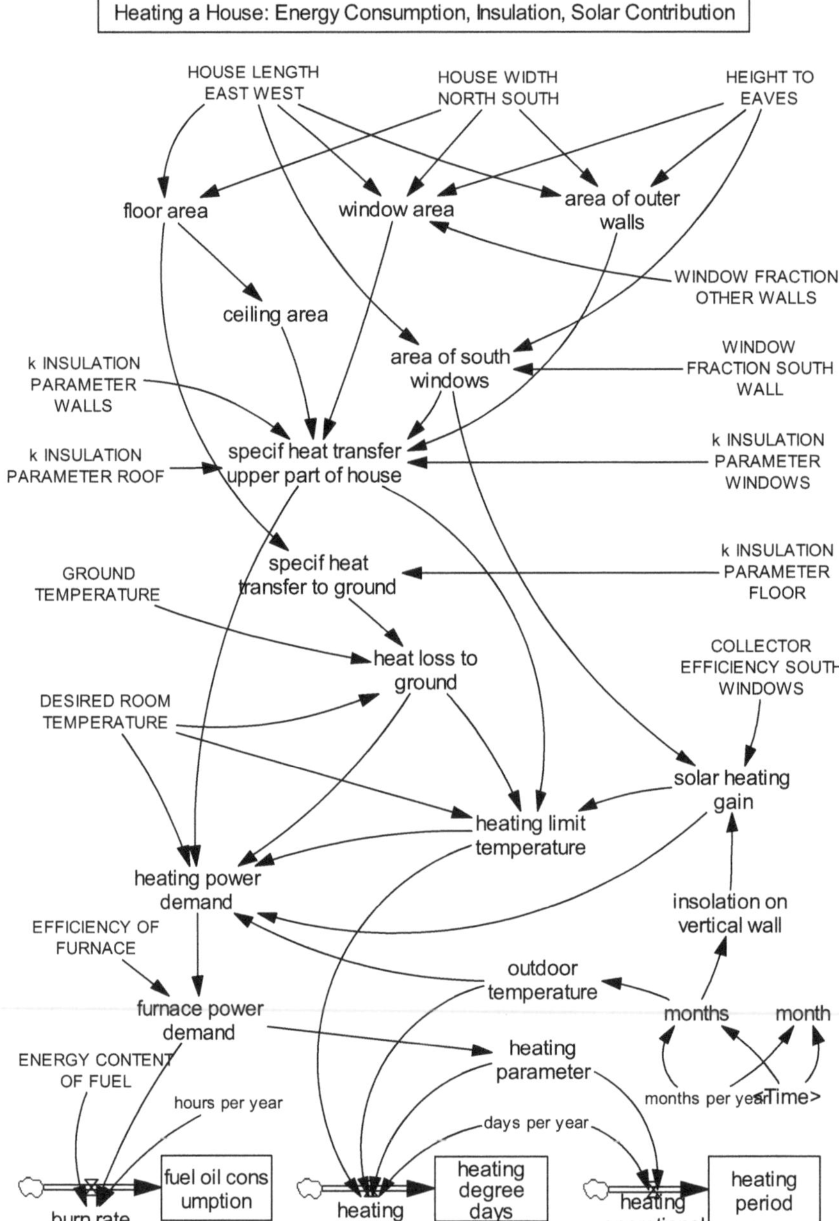

Figure Z212b: Simulation diagram for house heating dynamics.

Building parameters

HOUSE LENGTH EAST WEST = 12.5 [m]
HOUSE WIDTH NORTH SOUTH = 8 [m]
HEIGHT TO EAVES = 3 [m]
WINDOW FRACTION SOUTH WALL = 0.8 [1]
WINDOW FRACTION OTHER WALLS = 0.2 [1]
area of outer walls = 2* HOUSE WIDTH NORTH SOUTH * HEIGHT TO EAVES +2* HOUSE
        LENGTH EAST WEST * HEIGHT TO EAVES [m²]
floor area = HOUSE WIDTH NORTH SOUTH * HOUSE LENGTH EAST WEST [m²]
ceiling area = floor area [m²]
area of south windows = WINDOW FRACTION SOUTH WALL * HOUSE LENGTH EAST WEST *
        HEIGHT TO EAVES [m²]
window area = WINDOW FRACTION OTHER WALLS *(2* HOUSE WIDTH NORTH SOUTH *
        HEIGHT TO EAVES + HOUSE LENGTH EAST WEST * HEIGHT TO EAVES) [m²]

Heat transfer parameters

k INSULATION PARAMETER WALLS = 0.15 [W/(m²*C)]
k INSULATION PARAMETER ROOF = 0.2 [W/(C*m²)]
k INSULATION PARAMETER FLOOR = 0.2 [W/(C*m²)]
k INSULATION PARAMETER WINDOWS = 1.6 [W/(C*m²)]
COLLECTOR EFFICIENCY SOUTH WINDOWS = 0.8 [1]

Insolation, room and outdoor temperatures

GROUND TEMPERATURE = 9 [C] {deg C, centigrade}
DESIRED ROOM TEMPERATURE = 19 [C]
insolation on vertical wall = WITH LOOKUP (months, ([(0, 0) -(12, 125)], (0, 102), (0.5,
        101), (1.5, 111), (2.5, 105), (3.5, 79), (4.5, 45), (5.5, 38), (6.5, 48), (7.5, 69),
        (8.5, 99), (9.5, 102), (10.5, 106), (11.5, 104), (12, 102))) [W/(m²)]
outdoor temperature = WITH LOOKUP (months, ([(0, 0) -(12, 20)], (0, 17), (0.5, 17.8),
        (1.5, 17.3), (2.5, 14.2), (3.5, 9.1), (4.5, 4.9), (5.5, 1.5), (6.5, 0), (7.5, 0.8), (8.5,
        4.6), (9.5, 8.8), (10.5, 13.2), (11.5, 16.4), (12, 17))) [C]

Furnace parameters

ENERGY CONTENT OF FUEL = 11900 [W*Hour/liter]
EFFICIENCY OF FURNACE = 0.85 [1]

Heat gains and losses

solar heating gain = area of south windows * COLLECTOR EFFICIENCY SOUTH WINDOWS
        *insolation on vertical wall [W]
specif heat transfer upper part of house = (k INSULATION PARAMETER WALLS *(area of
        outer walls -(window area +area of south windows))) + k INSULATION PARAMETER
        ROOF *ceiling area + k INSULATION PARAMETER WINDOWS *(window area +area of
        south windows) [W/C]
specif heat transfer to ground = k INSULATION PARAMETER FLOOR *floor area [W/C]
heat loss to ground = specif heat transfer to ground *(DESIRED ROOM TEMPERATURE -
        GROUND TEMPERATURE) [W]

Heating process

heating limit temperature = DESIRED ROOM TEMPERATURE +(heat loss to ground -solar
        heating gain) / specif heat transfer upper part of house [C]
heating = (heating limit temperature -outdoor temperature) *days per year *heating
        parameter [C*Day/Year]

heating power demand = IF THEN ELSE (heating limit temperature <= outdoor temperature, 0, specif heat transfer upper part of house *(DESIRED ROOM TEMPERATURE -
outdoor temperature) +heat loss to ground -solar heating gain) [W]
furnace power demand = heating power demand / EFFICIENCY OF FURNACE [W]
heating parameter = IF THEN ELSE (furnace power demand > 0, 1, 0) [1]
heating operational = heating parameter *days per year [Day/Year]
fuel oil consumption = INTEG (burn rate, 0) [liter]
heating degree days = INTEG (heating, 0) [Day*C]
heating period = INTEG (heating operational, 0) [Day]
burn rate = (hours per year /(ENERGY CONTENT OF FUEL)) *furnace power demand [li-
ter/Year]

Time relationships
month = (Time*months per year) [Month]
months = (Time -INTEGER(Time)) *months per year [Month]
months per year = 12 [Month/Year]
days per year = 365 [Day/Year]
hours per year = 8760 [Hour/Year]

Simulation time parameters
INITIAL TIME = 0 [Year]
FINAL TIME = 1 [Year]
SAVEPER = TIME STEP [Year]
TIME STEP = 0.02 [Year]

Simulation results

In Figure Z212c simulation results are shown for the heating dynamics over a full year
of a "normal house" with average insulation values ("standard building"). By comparison, Figure Z212d shows the corresponding time plots for an "energy efficient
house" having very good insulation and large south-facing windows. In these diagrams the heating period (winter) corresponds to the center of the plot. The following
parameter values are used in the simulations:

Parameter	Standard building	Energy-efficient building
length east-west [m]	10	12.5
width north-south [m]	10	8
height to eaves [m]	3	3
window fraction south wall [%]	30	80
window fraction other walls [%]	25	20
insulation param walls [W/m^2 K]	1.3	0.15
insulation param roof [W/m^2 K]	1	0.2
insulation param floor [W/m^2 K]	0.9	0.2
insulation param windows [W/m^2 K]	4	1.6
room temperature [C]	19	19

On monthly average, heating is required for the "normal house" during the en-
tire year to maintain a room temperature of 19 degrees centigrade. In January the *furnace power demand* reaches a maximum of more than 8000 W. Total *fuel oil consumption* amounts to about 3200 liters per year.

The simulations produce the following results for one year:

Results	Standard building	Energy-efficient building
fuel oil consumption [Liter/year]	3196	220
heating degree days [deg C days]	3950	868
heating period [days]	365	119

By contrast, the "energy efficient house" must be heated only during a few months, and then only with a maximum *furnace power demand* of about 1400 W in January. Accordingly, the total *fuel oil consumption* amounts to only about 220 liters per year. The main reason for this considerable energy saving in comparison with the "normal house" is the very good insulation of all outer walls.

The following table provides an overview of the connection between wall type and heat loss (k-value). It can also be used for further simulation experiments.

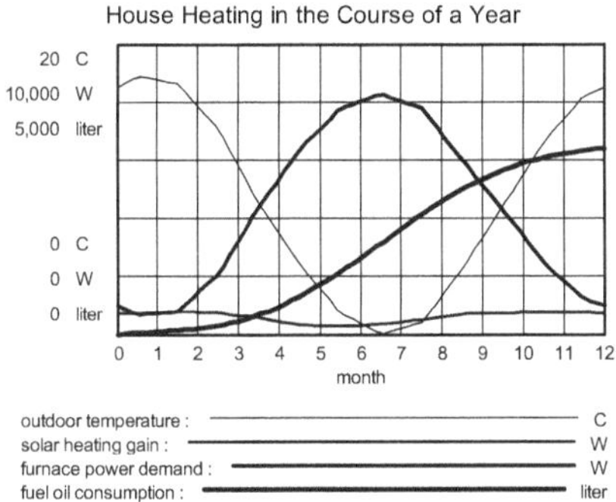

Figure Z212c: Heating variables for a "normal house" with poor insulation.

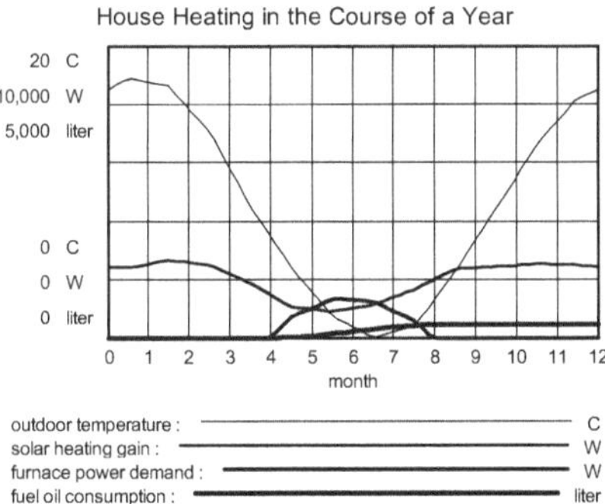

Figure Z212d: Heating variables for an "energy efficient house" with good insulation.

Heat transfer parameters (k-value) [W/(m² h)]

	good	buildings today	poor
outer walls	0.15	1.3	1.6
roof	0.2	1.0	1.2
floor	0.2	0.9	1.1
windows	1.6	4.0	5.7

Exercises

1. Experiment with different values for the insulation of windows and walls (k-values), for desired room temperature, and for size and placement of window areas particularly in the south-facing wall. In this manner, obtain an overview of the spectrum of the possible energy consumption values providing identical energy service to residents.

2. Examine systematically the consequences of better insulation of all components on total *fuel oil consumption*, on the number of *heating degree days*, and on the length of the *heating period* (for identical house geometry and desired room temperature).

3. Examine systematically the consequences of different house geometries (e.g. normal house, large south windows, earth house, wide or narrow south front, etc.).

4. Investigate the consequences of different *desired room temperatures*.

5. Use different heating systems (natural gas, wood pellets, condensing boiler, heat pump etc.) for heating. Change the EFFICIENCY OF FURNACE accordingly.

6. Complement the model e.g. by adding air exchange losses (leaks in window and door frames). How do the results change particularly for very well insulated houses?

References

Feist, W. 1999: *Das Passivhaus*. C. F. Müller, Heidelberg.

Krause, F., Bossel, H., Müller-Reißmann, K.F. 1980: *Energie-Wende – Wachstum und Wohlstand ohne Erdöl und Uran*. Fischer, Frankfurt/M.

H. Bossel 1985: *Umweltdynamik – 30 Programme für kybernetische Umwelterfahrungen auf jedem BASIC-Rechner*. Te-wi Verlag, München (321-334).

Z213 Integral relations and heat conduction

Simulation task

So far we have exclusively dealt with systems whose state variables were functions of time, and independent of location in space. The system can then be described by ordinary differential equations. In this case only derivatives of state variables with respect to time appear in the equations.

Not all systems of practical interest can be modeled in this way to simulate their dynamics. There is a broad class of systems where the spatial distribution of state variables is an essential feature. In this case gradients of state variables exist between neighboring points, which have to be described by additional derivatives, now with respect to the spatial coordinates. If derivatives with respect to more than one independent coordinate have to be included, one speaks of *partial* differential equations. In the examples so far only time was involved, and we were dealing with *ordinary* differential equations.

Systems where the spatial distribution of state variables is a crucial and elementary feature are, for example, airflow over a wing, pressure and temperature distribution in weather patterns, diffusion of pollutants in atmosphere or groundwater, distribution of stresses in a load bearing shell, or heat diffusion in a machine part.

In principle such systems can be approximated by a model where the space in question is filled with a large number of grid points. One or several state variables are defined at each of these points, and their dependence on conditions at neighboring points and on time is computed using the locally applicable partial differential equations. The method of finite differences employs this approach.

It is obvious that even a rather coarse system of grid points requires computation of a large number of (locally expressed) state variables, thus taxing speed and storage capabilities of even large computers. For this reason, finite difference computations of accurate and long term weather forecasts are very demanding in terms of computer capacity and performance.

In many cases, the efficiency of the numerical solution of partial differential equations can be improved by making use of the fact that the spatial distribution of state variables must obey certain conditions (initial conditions, boundary conditions, continuity conditions) and is therefore not completely arbitrary. By using this knowledge, appropriate approximations can be applied which lead to more efficient computing procedures. The method of finite elements in particular, which is based on such knowledge, has found wide application for this reason.

In the following, the related method of integral relations (also: method of weighted residuals, or Ritz-Galerkin method) is described and used. Using this method it is possible to reduce many partial differential equations in two coordinates to a system of ordinary differential equations in one coordinate (e.g. time). The procedure is employed, for example, for the computation of boundary layer flows or vortex and wake flows. We apply it here in model Z213 to the partial differential equation of diffusion, more exactly to time-dependent thermal conduction in an insulated bar. In model Z214 the procedure is also used for the more complicated task of computing boundary layer flow.

The method of integral relations

We look in the following at a procedure for the solution of partial differential equations in two coordinates for a field variable (distributed variable) $u(x, y)$ depending on two coordinates x and y, where one of these independent variables can be time t. Both the method of integral relations and the method of finite differences solve a corresponding partial differential equation numerically by approximation of the dependent variable, here $u(x, y)$. The first procedure uses (at least partially) continuous approximation functions, the second procedure employs discretization at the grid points of the computation and approximation of the partial differentials by expressions containing the function values at adjacent grid points. The solution requires determining the parameters of the approximation.

In the method of finite differences the predefined functions $\delta(y_n)$ are Dirac delta functions and the $u_n(x)$ are the parameter solutions to be determined.

$$u(x, y) \approx u^*(x, y) = \sum_{n=1}^{N} u_n(x) \cdot \delta(y_n)$$

The method of finite differences requires far more parameters for an exact solution than the method of integral relations, which therefore has an advantage with respect to computing time. However, the formulation of the integral relations is more complicated than the formulation of the expressions for the method of finite differences.

Before getting into concrete applications later, the steps required for formulating the method of integral relations shall be described. Let it be required to solve a (nonlinear or linear) partial differential equation symbolically represented by

$$L(u) = 0$$

This equation is to be solved for the dependent field variable $u(x, y)$ while respecting the boundary conditions valid at the borders of the integration range of interest. We develop the approach here particularly for the partial differential equations of diffusion, and of boundary layer flows and other flows of parabolic character (initial value problem), where an initial profile $u(a, y)$ and boundary conditions $u(x, c)$ (at the lower edge) and $u(x, d)$ (at the upper edge) are prescribed. The integration then runs from $x = a$ in the direction of increasing x (Figure Z213a).

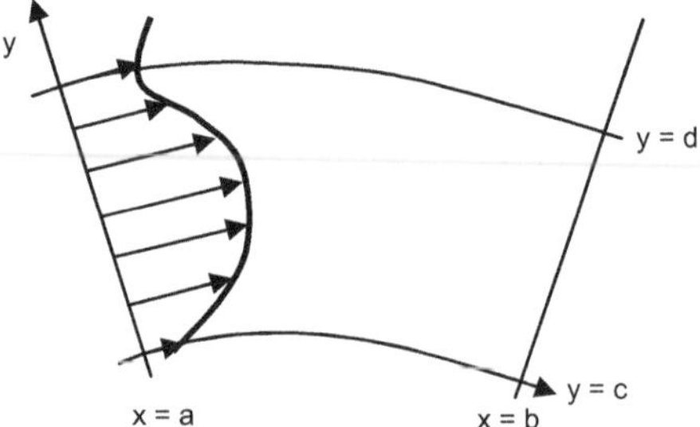

Figure Z213a: Application of the method of integral relations to a system of (parabolic) partial differential equations in two independent coordinates x and y. Approximation of the solution function in one dimension (y) by parameters $a(x)$ and numerical integration in the other dimension (x). The initial profile $u(x, y)$ at $x = a$ and the boundary conditions at $y = c$ und $y = d$ are prescribed.

In the method of integral relations the unknown solution $u(x, y)$ is approximated by an approximation function $u^*(x, y)$. It contains predefined functions f_n in one independent variable (y for boundary layer problems) which is multiplied by initially unknown parameters a_n that are functions of the other independent variables (here: x).

$$u(x, y) \approx u^*(x, y) = \sum_{n=1}^{N} a_n(x) \cdot f_n(y)$$

The $f_n(y)$ are usually selected such that they satisfy the boundary conditions $u(x, c)$ and $u(x, d)$. If this approximating expression is entered into the partial differential equation with initially arbitrary parameters a, the equation is generally not satisfied at first and an error (residual) arises:

$$L(u^*) = R(x, y) \neq 0$$

The n parameters $a_n(x)$ are now determined by applying the condition that the n integrals of this residual over the region of integration in the y-direction must disappear (i.e. be zero). To obtain a set of linearly independent equations for the $a_n(x)$, the residual function $R(x, y)$ must be multiplied before integration by elements of a set of linearly independent weighting functions $w_k(y)$, i.e.

$$\int_c^d w_k\, R(x, y)\, dy = 0 , \quad k = 1, 2, \cdots, N$$

This procedure therefore corresponds to multiplication of the partial differential equation $L(u) = 0$ by weighting functions $w_k(y)$, formal integration with respect to y and replacement of $u(x, y)$ in the integrals by $u^*(x, y)$, to allow integration and to thereby eliminate the y-dependence.

$$\int_c^d w_k(y) \cdot L\{u^*[a_n(x), f_n(y)]\}\, dy = 0$$

After this integration a set of N ordinary differential equations in x remains for the $a_n(x)$. These ordinary differential equations can then be integrated numerically in the x-direction with usual methods (Euler-Cauchy, Runge-Kutta or predictor-corrector methods). Inserting the $a_n(x)$ computed in this manner into the approximation expression provides the solution of the task.

The success of a specific application of the method of integral relations depends critically on the choice of the approximation functions $f_n(y)$ and partly also on the choice of the weighting functions $w_k(y)$. Correctly formulated implementations show convergence to the exact solution if the number of parameters a_n is increased. The use of few approximation terms ($N = 1$ or 2) generally leads to good (engineering) estimates, while for $N = 3$ and higher very exact solutions can be obtained.

Differential equations of heat conduction in a bar

The arrangement is outlined in Figure Z213b. A (metal) bar of constant initial temperature, perfectly insulated on all sides except at its two ends, is heated (or cooled) at both ends. The time-dependent temperatures at both ends are prescribed. The temperature distribution at all locations X in the bar is to be computed as function of time.

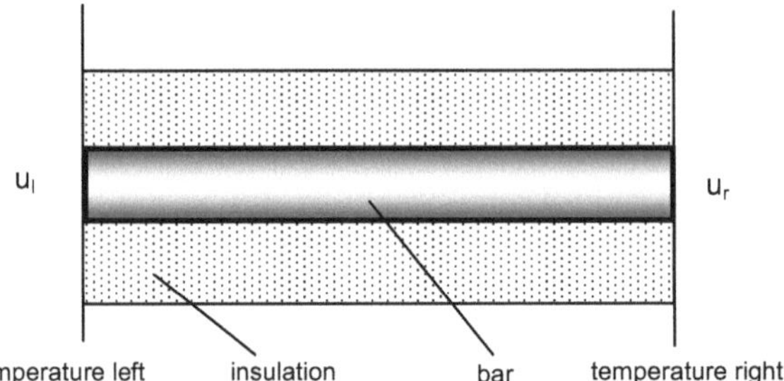

Figure Z213b: Insulated bar with prescribed (time-dependent) temperature at both ends.

The one-dimensional equation of heat diffusion applies to this problem. This partial differential equation contains the first derivative of temperature with respect to time t [h] and the second derivative of temperature with respect to location X [m]. At each position in the bar, a different space and time dependent temperature distribution develops. Heat conduction depends on the heat conduction coefficient α [m²/h]. The diffusion equation is written as

$$\frac{\partial u}{\partial t} = \alpha \frac{\partial^2 u}{\partial X^2} \tag{1}$$

with initial condition

$$u(X,0) = u_a$$

and the boundary conditions (left and right, length of bar L [m])

$$u(0,t) = u_l(t)$$
$$u(L,t) = u_r(t)$$

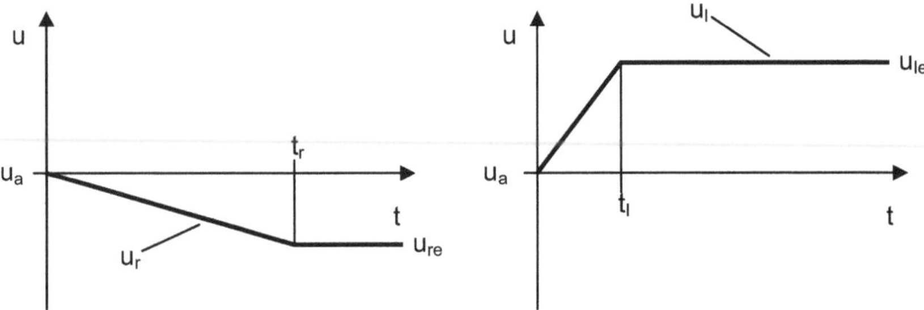

Figure Z213c: Parameters of the time-dependent temperatures at the right hand (subscript *r*) and left hand end (subscript *l*).

The temperature u is expressed as relative (nondimensional) quantity. It is related to real temperature T [K] by the following relationship:

$$u = \frac{T - T_0}{T_1 - T_0} = \frac{\Delta T}{\Delta T_{ges}}$$

The dimensionless temperature u corresponds to the ratio of the current temperature difference ΔT to the maximum temperature difference, where

T = current (local) temperature [K]
T_0 = minimal temperature of the bar [K]
T_1 = maximum temperature [K] =

It is convenient to work with nondimensional quantities throughout. A nondimensional coordinate x can be defined as

$$x = \frac{X}{L}$$

Here X is the real coordinate [m] and L = bar length [m]. In the expression for nondimensional time

$$\tau = \frac{\alpha t}{L^2}$$

t = real time [h] and α = coefficient of heat conduction [m²/h]. Correspondingly, real time t is given by

$$t = \frac{L^2 \tau}{\alpha}$$

The coefficient of heat conduction differs significantly for different materials (Hütte 1955: 495).

Material	coefficient of heat conduction [m²/h]
copper	0.4
steel	0.04
concrete	0.002
cork	0.0005

Replacing t by τ and X by x the nondimensional form of the heat diffusion equation is obtained. It will be used in the further work.

$$\frac{\partial u}{\partial \tau} = \frac{\partial^2 u}{\partial x^2} \tag{1'}$$

The coefficient of heat conduction no longer appears in the equation as a parameter, but it still has a decisive influence on the result for the real variables, since these are determined by conversion from the nondimensional variables. This applies in particular to the coefficient of heat conduction. If it is small (poor thermal conduction), then the real time is "stretched" accordingly: The diffusion process then develops accordingly more slowly.

Model formulation

For modeling unsteady thermal conduction in a bar we follow the general approach discussed above. Details of the derivation of the integral relations are found elsewhere (Bossel 1992: 273-277).

(1) First, the partial differential equation (here: diffusion equation) is multiplied by a general weighting function f_k and formally integrated with respect to the x-coordinate (spatial coordinate). Before the resulting equation can be processed further, suitable formulations must be chosen for the approximation of the solution $u(x, \tau)$ and the weighting functions $f_k(x)$.

(2) For the approximation function an expression is selected that already satisfies the boundary conditions at both ends of the bar automatically, and that furthermore facilitates the representation of the temperature profiles to be expected. Obviously a Fourier approximation with factors $a_k(\tau)$, and sine components which disappear at both ends is a suitable candidate for this application. We therefore select, with time-dependent boundary conditions $u_r(\tau)$ and $u_l(\tau)$ at the right and left end of the bar

$$u(x,\tau) = u_l(\tau) + (u_r - u_l)x + \sum_{n=1}^{N} a_n(\tau) \cdot \sin n\pi x$$

(3) The weighting functions $f_k(x)$ must represent elements from a set of linearly independent functions which in addition should lead to integral expressions that are as simple as possible. The Fourier representation chosen for the approximation suggests the use of sine functions also as weighting functions.

$$f_k(x) = \sin k\pi x$$

(4) If now the chosen approximation and weighting functions are inserted in the integral relations, the x-dependence disappears after integration with respect to x, and a system of ordinary differential equations with time as the only independent variable remains.

(5) The integral expressions can be considerably simplified considering the applicable integration rules. A system of k differential equations for the a_k remains, which can now be computed at every instant as a function of temperature conditions at the left and right hand ends and their first derivative with respect to time.

(6) The rates of change of the state variables a_k can be integrated numerically with usual procedures. By introducing the resulting parameters $a_k(\tau)$ into the approximation function for the space and time-dependent temperature distribution $u(x, \tau)$, the space and time-dependent solution of the problem is obtained.

For the rates of change $\dot{a}_k$ of the approximation parameters a_k the following ordinary differential equations are obtained by this procedure:

$$\dot{a}_k = \frac{da_k}{dt}$$

$$= 2\left[\frac{\dot{u}_r}{k\pi}\cos k\pi - \frac{\dot{u}_l}{k\pi} + \left\{(u_l - u_r \cos k\pi)k\pi - \pi^2 k^2\left[\frac{u_l}{k\pi} - \frac{u_r}{k\pi}\cos k\pi + \frac{a_k}{2}\right]\right\}\right]$$

Here the boundary conditions u_l and u_r for the temperature functions on the right and on the left, and their first derivatives with respect to time must be provided. We choose a linear change from the initial temperature of the bar to the final temperature on the left (u_{le}) and on the right (u_{re}). The final temperature is reached on the left after time τ_l, on the right after time τ_r (Figure Z213c). The resulting boundary conditions are on the left:

$$\dot{u}_l = \frac{u_{le} - u_a}{t_l}\,, \quad u_l = u_a + \dot{u}_l\,\tau \quad for \ 0 \leq \tau < \tau_l$$

$$\dot{u}_l = 0 \qquad , \quad u_l = u_{le} \qquad for \ \tau > \tau_l$$

on the right:

$$\dot{u}_r = \frac{u_{re} - u_a}{t_r}\,, \quad u_r = u_a + \dot{u}_r\,\tau \quad for \ 0 \leq \tau < \tau_r$$

$$\dot{u}_r = 0 \qquad , \quad u_r = u_{re} \qquad for \ \tau > \tau_r$$

The ordinary differential equations for the a_k are solved numerically with these boundary conditions and the initial condition

$$u(x,0) = u_a$$

Simulation model

The program statements are listed in the following (normal type for $N = 3$, smaller type for the additional statements for $N = 9$). The corresponding simulation diagrams for three approximation terms ($N = 3$) are shown in Figures Z213d and e. If more approximation terms are to be used, the simulation diagrams have to be completed by analogous structures.

The parameters LENGTH OF BAR, (relative) INITIAL TEMPERATURE OF BAR and HEAT CONDUCTION PARAMETER (0.1 for brass) have to be entered first. Next, the (relative) FINAL TEMPERATURE at both ends of the bar as well as the REAL TIME FOR FINAL TEMPERATURE are specified. These times are converted to dimensionless time (*time* [1]) as used in the simulation. The temperature functions at the left (UL) and right (UR) hand ends of the bar are defined as linear functions. The corresponding slopes (DUL and DUR) are calculated. With these data, which prescribe the temporal development of the boundary conditions, the rates of change D_k of the approximation parameters A_k can be computed. The A_k follow by numerical integration over dimensionless *time*. For representation of the results in *real time*, the simulation *time* must be converted to *real time* by multiplication with (LENGTH OF ROD^2/HEAT CONDUCTION PARAMETER).

After inserting the computed approximation parameters $A_k = a_k(t)$ into the approximation formula the temperature distribution $u(x, t)$ in the bar can be calculated as a function of time. If the time series for the parameter A_k are transferred to a spreadsheet program, the results can be displayed in various ways, in particular in a three-dimensional presentation of $u(x, t)$.

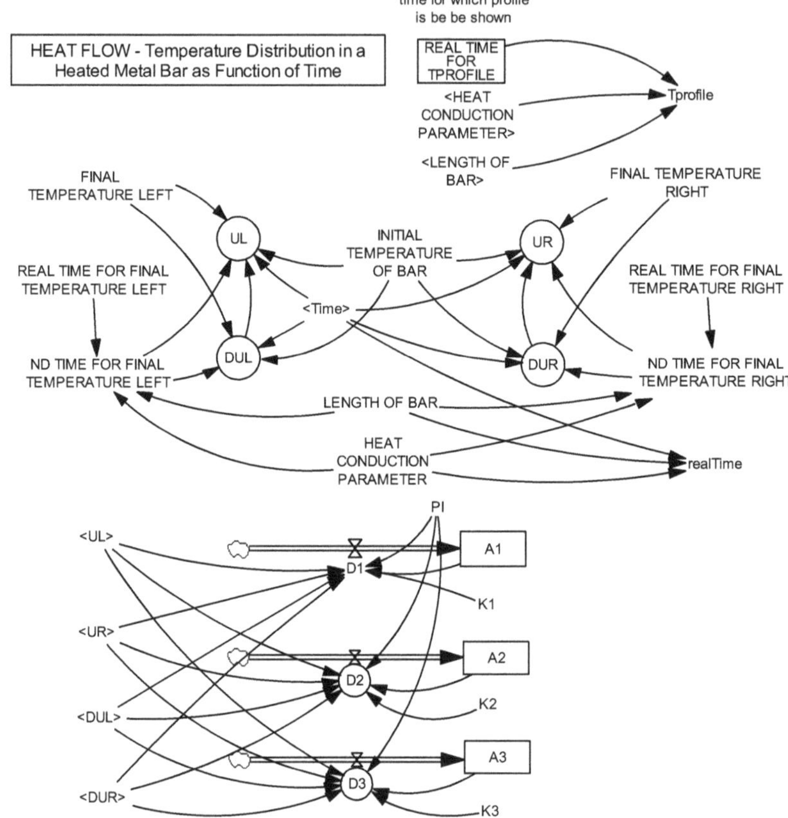

Figure Z213d: Simulation diagram for computation of time-dependent heat conduction in a bar using the method of integral relations – Part 1.

The program computes the velocity profile for a specified point in *time* (REAL-TIME FOR TPROFILE). The corresponding program statements are found under the heading "Temperature profile for a given point in time" below. When (computation) time reaches the specified TPROFILE, further time steps are redefined as steps in the *x*-direction to compute the lengthwise temperature profile in the bar at this point in time. This profile can then be plotted as *TempProfile(x)*.

Constants and parameters
PI = 3.14159 [1]
LENGTH OF BAR = 1 [m]
HEAT CONDUCTION PARAMETER = 0.1 [m*m/Hour]
REAL TIME FOR TPROFILE = 1

Initial and boundary conditions
INITIAL TEMPERATURE OF BAR = 0 [1]
        {dimensionless temperature between 0 and 1}

FINAL TEMPERATURE LEFT = 1 [1] {dimensionless temperature}
realTime FOR FINAL TEMPERATURE LEFT = 0.2 [Hour]
NDtime FOR FINAL TEMPERATURE LEFT = realTime FOR FINAL TEMPERATURE
    LEFT *HEAT CONDUCTION PARAMETER /LENGTH OF BAR^2 [1]
    {dimensionless time}
FINAL TEMPERATURE RIGHT = 1 [1] {dimensionless temperature}
realTime FOR FINAL TEMPERATURE RIGHT = 0.8 [Hour]
NDtime FOR FINAL TEMPERATURE RIGHT = realTime FOR FINAL TEMPERA-
    TURE RIGHT *HEAT CONDUCTION PARAMETER /LENGTH OF BAR^2 [1]
    {dimensionless time}
UL = IF THEN ELSE (Time > NDtime FOR FINAL TEMPERATURE LEFT, FINAL
    TEMPERATURE LEFT, INITIAL TEMPERATURE OF BAR +Time *DUL) [1]
UR = IF THEN ELSE (Time > NDtime FOR FINAL TEMPERATURE RIGHT, FINAL
    TEMPERATURE RIGHT, INITIAL TEMPERATURE OF BAR +Time *DUR) [1]
DUL = IF THEN ELSE (Time > NDtime FOR FINAL TEMPERATURE LEFT, 0, (FINAL
    TEMPERATURE LEFT -INITIAL TEMPERATURE OF BAR) /NDtime FOR FI-
    NAL TEMPERATURE LEFT) [1]
DUR = IF THEN ELSE (Time > NDtime FOR FINAL TEMPERATURE RIGHT, 0, (FI-
    NAL TEMPERATURE RIGHT -INITIAL TEMPERATURE OF BAR) /NDtime
    FOR FINAL TEMPERATURE RIGHT) [1]

Time conversion
realTime = Time*LENGTH OF BAR^2/HEAT CONDUCTION PARAMETER [Hour]

Coefficients
K1 = 1 [1]
K2 = 2 [1]
K3 = 3 [1]
K4 = 4 [1]
K5 = 5 [1]
K6 = 6 [1]
K7 = 7 [1]
K8 = 8 [1]
K9 = 9 [1]

Rates of change of approximation parameters
D1 = 2*(DUR*COS(K1*PI) /(K1*PI) -DUL/(K1*PI) +((UL-UR*COS(K1*PI)) *K1*PI-
    PI*PI*K1* K1*(UL/(K1*PI) -UR*COS(K1*PI) /(K1*PI) +A1/2))) [1]
D2 = 2*(DUR*COS(K2*PI) /(K2*PI) -DUL/(K2*PI) +((UL-UR*COS(K2*PI)) *K2*PI-
    PI*PI*K2* K2*(UL/(K2*PI) -UR*COS(K2*PI) /(K2*PI) +A2/2))) [1]
D3 = 2*(DUR*COS(K3*PI) /(K3*PI) -DUL/(K3*PI) +((UL-UR*COS(K3*PI)) *K3*PI-
    PI*PI*K3* K3*(UL/(K3*PI) -UR*COS(K3*PI) /(K3*PI) +A3/2))) [1]
D4 = 2*(DUR*COS(K4*PI) /(K4*PI) -DUL/(K4*PI) +((UL-UR*COS(K4*PI)) *K4*PI-PI*PI*K4* K4*(UL/(K4*PI) -UR*COS(K4*PI) /(K4*PI) +A4/2)))
    [1]
D5 = 2*(DUR*COS(K5*PI) /(K5*PI) -DUL/(K5*PI) +((UL-UR*COS(K5*PI)) *K5*PI-PI*PI*K5* K5*(UL/(K5*PI) -UR*COS(K5*PI) /(K5*PI) +A5/2)))
    [1]
D6 = 2*(DUR*COS(K6*PI) /(K6*PI) -DUL/(K6*PI) +((UL-UR*COS(K6*PI)) *K6*PI-PI*PI*K6* K6*(UL/(K6*PI) -UR*COS(K6*PI) /(K6*PI) +A6/2)))
    [1]
D7 = 2*(DUR*COS(K7*PI) /(K7*PI) -DUL/(K7*PI) +((UL-UR*COS(K7*PI)) *K7*PI-PI*PI*K7* K7*(UL/(K7*PI) -UR*COS(K7*PI) /(K7*PI) +A7/2)))
    [1]
D8 = 2*(DUR*COS(K8*PI) /(K8*PI) -DUL/(K8*PI) +((UL-UR*COS(K8*PI)) *K8*PI-PI*PI*K8* K8*(UL/(K8*PI) -UR*COS(K8*PI) /(K8*PI) +A8/2)))
    [1]
D9 = 2*(DUR*COS(K9*PI) /(K9*PI) -DUL/(K9*PI) +((UL-UR*COS(K9*PI)) *K9*PI-PI*PI*K9* K9*(UL/(K9*PI) -UR*COS(K9*PI) /(K9*PI) +A9/2)))
    [1]

Approximation parameters as integrals of dimensionless time (Time)
A1 = INTEG(D1, 0) [1]
A2 = INTEG(D2, 0) [1]
A3 = INTEG(D3, 0) [1]
A4 = INTEG(D4, 0) [1]
A5 = INTEG(D5, 0) [1]
A6 = INTEG(D6, 0) [1]
A7 = INTEG(D7, 0) [1]
A8 = INTEG(D8, 0) [1]
A9 = INTEG(D9, 0) [1]

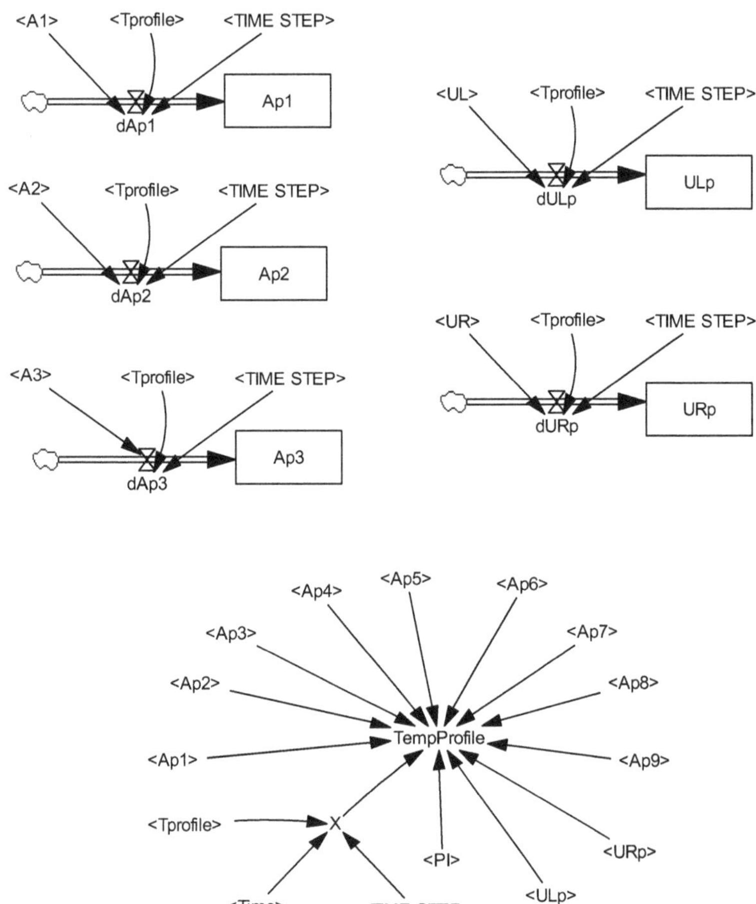

Figure Z213e: Simulation diagram for computation of time-dependent heat conduction in a bar – Part 2: Calculation of temperature profile.

Temperature profile for a given point in time
Tprofile = REAL TIME FOR TPROFILE *HEAT CONDUCTION PARAMETER
/LENGTH OF BAR^2
Rates of change of profile parameters:
dAp1 = A1 *PULSE(Tprofile, TIME STEP) /TIME STEP
dAp2 = A2 *PULSE(Tprofile, TIME STEP) /TIME STEP
dAp3 = A3 *PULSE(Tprofile, TIME STEP) /TIME STEP
dAp4 = A4 *PULSE(Tprofile, TIME STEP) /TIME STEP
dAp5 = A5 *PULSE(Tprofile, TIME STEP) /TIME STEP
dAp6 = A6 *PULSE(Tprofile, TIME STEP) /TIME STEP
dAp7 = A7 *PULSE(Tprofile, TIME STEP) /TIME STEP
dAp8 = A8 *PULSE(Tprofile, TIME STEP) /TIME STEP
dAp9 = A9 *PULSE(Tprofile, TIME STEP) /TIME STEP

Profile parameters:
Ap1 = INTEG (dAp1, 0)
Ap2 = INTEG (dAp2, 0)
Ap3 = INTEG (dAp3, 0)
Ap4 = INTEG (dAp4, 0)
Ap5 = INTEG (dAp5, 0)
Ap6 = INTEG (dAp6, 0)
Ap7 = INTEG (dAp7, 0)
Ap8 = INTEG (dAp8, 0)
Ap9 = INTEG (dAp9, 0)

Boundary conditions:
dULp = UL *PULSE(Tprofile, TIME STEP) /TIME STEP
dURp = UR *PULSE(Tprofile, TIME STEP) /TIME STEP
ULp = INTEG (dULp, 0)
URp = INTEG (dURp, 0)

Temperature profile for time = Tprofile:
X = IF THEN ELSE(Time >= Tprofile :AND: Time <= Tprofile +100 *TIME STEP,
    (Time-Tprofile) /(100 *TIME STEP), 0)

Temperature profile as f(X):
TempProfile = ULp+(URp -ULp) *X
    +Ap1 *SIN(PI *X) +Ap2 *SIN(2 *PI *X) +Ap3 *SIN(3 *PI *X)
    +Ap4 *SIN(4 *PI *X) +Ap5 *SIN(5 *PI *X) +Ap6 *SIN(6 *PI *X)
    +Ap7 *SIN(7 *PI *X) +Ap8 *SIN(8 *PI *X) +Ap9 *SIN(9 *PI *X)

Simulation time parameters
INITIAL TIME = 0 [1]
FINAL TIME = 0.2 [1]
TIME STEP = 0.001 [1]
SAVEPER = TIME STEP [1]

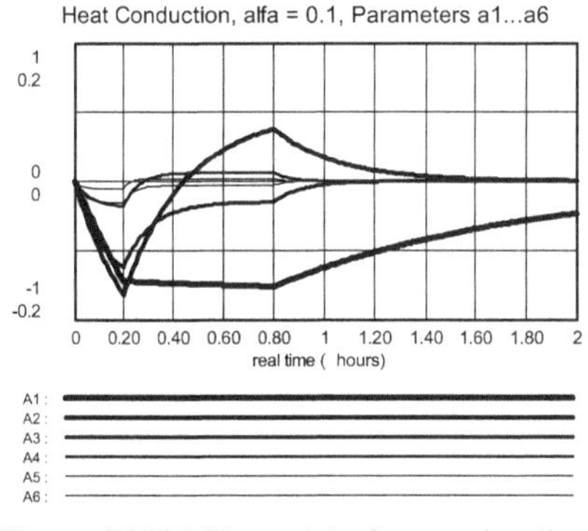

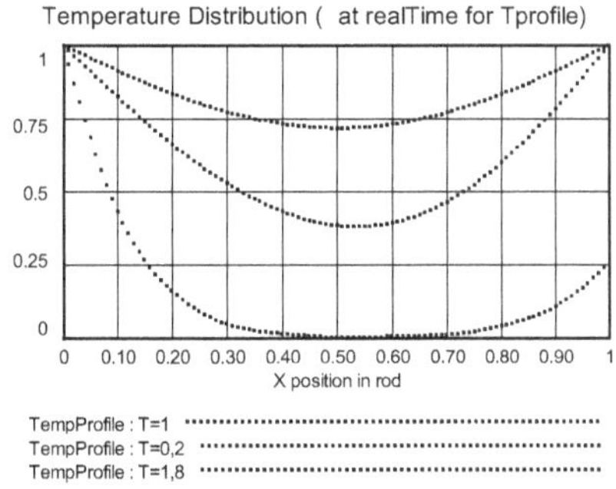

Figure Z213f: Time plot of approximation parameters for heat conduction in a bar.
Figure Z213g: Temperature distribution in the bar for (real) time *T* = 0.2, 1.0, 1.8.

Simulation results

The results of an exemplary simulation with $N = 9$ are shown in Figures Z213f and Z213g. In this case the left end of the bar was heated quickly, the right end only slowly. The first figure shows the time plots of the first 6 approximation parameters a_k, the second is a plot of temperature distribution in the bar at three different times after heating of both ends began. A three-dimensional representation of these results (after computation of $u(x, t)$ in a spreadsheet program) is provided in Figure Z213h. The picture clearly shows how the heat at both ends slowly diffuses towards the center until the bar finally assumes the new temperature. Figure Z213i shows results of a second simulation with heating of the left end and cooling of the right.

The program can be conveniently used to investigate approximation quality and convergence as function of the number of approximation terms. Even a small number of approximation terms (two or three) already produces acceptable results.

Exercises

1. Vary the initial temperature of the bar in the range from 0 to 1 and choose different rapid and strong changes of the final temperatures on the right and on the left. Run and document several such simulations. Use the same number of approximation members (e.g. 4 or 6) for all simulations.
2. Study the time plots for the N approximation parameters a_k. Explain how their pattern with time depends on the chosen boundary conditions. How do the absolute values of the approximation parameters relate to the order (number k) of the respective parameter? Can you infer convergence of the results from these observations?
3. Formulate and/or obtain numerical solutions of the partial differential equation using the method of finite differences. Apply the method of integral relations and the method of finite differences to identical initial and boundary conditions. Compare the results and the respective programming effort.
4. Why do the approximation parameters with an even index disappear for symmetric boundary conditions? Under what conditions would the approximation parameters with an odd index disappear?

References

Akademischer Verein Hütte (Hg.) 1955: *Hütte – Des Ingenieurs Taschenbuch*. Verlag Wilhelm Ernst, Berlin, 28th ed.
Bossel, H. 1992: *Simulation dynamischer Systeme – Grundwissen, Methoden, Programme*. Vieweg Verlag, Braunschweig und Wiesbaden, 2. Aufl.

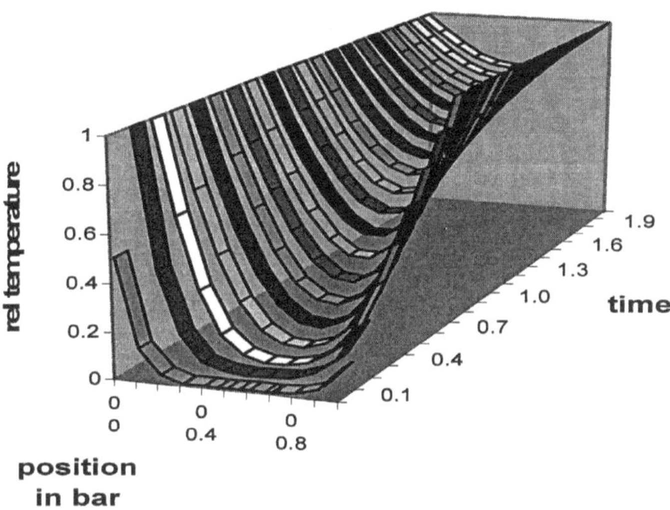

Figure Z213h: Temperature distribution in the bar as function of time: slow heating on the right, fast heating on the left.

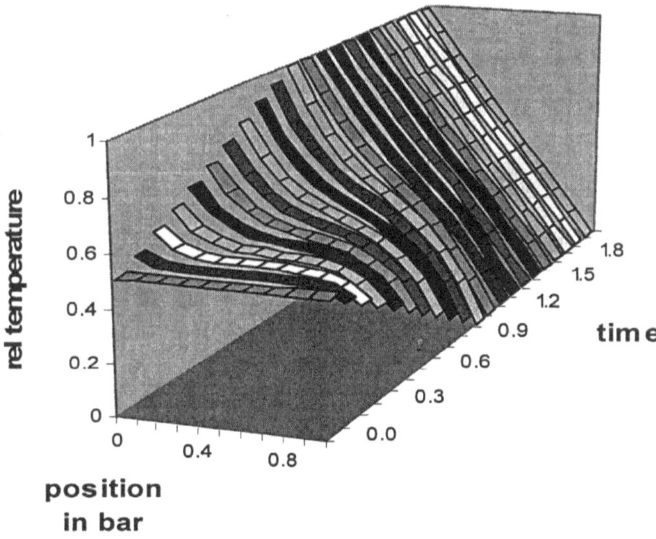

Figure Z213i: Temperature distribution in the bar as function of time: cooling on the right, heating on the left.

Z214 Boundary layer flow

Simulation task

To increase the energy efficiency of airplanes or improve the gliding performance of sailplanes, wing profiles which create the necessary lift with minimum drag must be developed. Optimal profiles for wings or other bodies are developed using corresponding computing programs. The computation is difficult and complex because viscous flows must be described with nonlinear partial differential equations, the Navier-Stokes equations. The computation must account for the fact that the flow at one point of a flow field also influences (in principle) the flow at every other point. This means that the whole flow field must be calculated simultaneously at all points – which requires a very significant computational effort, in particular for large scale flow fields like the atmosphere. Even supercomputers take days to compute the velocity, temperature, and pressure fields as functions of time for long-term global weather forecasts and climate computations.

Under certain conditions which are of practical importance in fluid dynamics the full flow equations can be simplified, and can then be computed with a far smaller effort. For example, the flow outside the "boundary layer" around a body can be reliably described by the differential equations for inviscid flow, because the viscous terms of the full flow equations become negligibly small. The inviscid equations are linear and can be solved by superposition of fundamental solutions (examples of the calculation of vortex flows and wind tunnel contractions: Bossel 1969a, 1969b). In the boundary layer itself, where by friction on the body surface the velocity of the external flow is reduced to zero at the surface, the full Navier-Stokes equations reduce to the nonlinear boundary layer equations. These are characterized by the fact that the downstream flow has no effect on the upstream flow. The flow field therefore must not be computed simultaneously at all positions. Instead, fluid flow in the boundary layer can be computed stepwise in the direction of flow. This corresponds to most simulations in this book, which use stepwise progression in time, without the future having any influence on the results for the present.

In two-dimensional boundary layer flow (e.g. around an airfoil) there still remains a system of nonlinear partial differential equations in two space coordinates for the velocity field $u(x, y)$. Coordinate x is in the direction of flow and (usually) has its origin at the stagnation point of the body. Coordinate y is orthogonal to x, has its origin (usually) at the surface of the body and extends to the outer edge of the boundary layer and the beginning of the free (inviscid) flow. Application of the method of integral relations developed in model Z213 "Heat conduction" allows reformulating the boundary layer equations in terms of a system of ordinary differential equations, which can be computed by stepwise progression in the x-direction. In this way it becomes possible to compute the *partial* differential equations of boundary layer flow with the same method used so far for the computation of time-dependent *ordinary* differential equations.

Formulating the integral relations

With the introduction of the dimensionless Reynolds number $Re = u_\infty \cdot l / v$ (with u_∞ [m/s] = velocity of the free flow, l [m] = characteristic length, v [m²/s] = kinematic viscosity) and the usual transformations

$$X = x/l, \ \ Y = Re^{1/2} \, y/l, \ \ U = u/u_\infty, \ \ V = Re^{1/2} \, v/u_\infty$$

the laminar incompressible boundary layer equations become nondimensional, and the momentum equation and continuity equation can be written as

$$\frac{\partial U^2}{\partial X} + \frac{\partial (VU)}{\partial Y} = U_e \frac{dU_e}{dX} + \frac{\partial^2 U}{\partial Y^2}$$

$$\frac{\partial U}{\partial X} + \frac{\partial V}{\partial Y} = 0$$

with the boundary conditions

$$U(X, 0) = 0, \ \ U(X, \infty) = U_e(X), \ \ V(X, 0) = V_0(X)$$
$$U(0, Y) = \text{initial profile}$$

The method of integral relations (method of the weighted residuals) is used to reduce the partial differential equations in X and Y to a set of ordinary differential equations in X. The Y-dependence is eliminated by assuming the functional form of the velocity profile $U(Y, a_n(X))$ in the Y direction, incorporating the correct behavior at the boundaries and employing N free parameters $a_n(X)$, and integrating the momentum equation in the Y-direction ($0 \leq Y \leq \infty$). The system of N linearly independent equations for the $a_n(X)$ is generated by multiplying the momentum equation by N members of a set of linearly independent weighting functions $f_k(Y)$ before integration. The integral relations for the laminar incompressible boundary layer flow then become

$$\frac{d}{dX} \int_0^\infty f_k U^2 \, dY - \int_0^\infty f'_k VU \, dY - \int_0^\infty f''_k U \, dY + \left[f_k \frac{\partial U}{\partial Y} \right]_{Y=0} - U_e \frac{dU_e}{dX} \int_0^\infty f_k \, dY = 0 \ \ ,$$

$$k = 1, 2, \cdots, N$$

Integration of the continuity equation permits replacement of V in this equation:

$$V(X,Y) = -\int_0^Y \frac{\partial U}{\partial X} dY + V_0(X)$$

Here $V_0(X)$ is the prescribed suction or blowing velocity (suction is negative).

To calculate the integrals, the weighting functions $f_k(Y)$ and the approximation function $U(Y, a_n(X))$ for the velocity U are chosen as follows:

$$f_k(Y) = e^{-\sigma_k Y} \, , \ k = 1, 2, \cdots, N$$

$$U(X,Y) = (1 - e^{-\alpha Y}) \cdot [U_e(X) + \sum_{n=1}^{N} a_n(X) e^{-n\alpha Y}]$$

Here α is a constant and should correspond roughly to the exponential decay of the velocity profile. The approximation for U satisfies the boundary conditions for $Y = 0$ und $Y \rightarrow \infty$.

Substitution of weighting functions $f_k(Y)$ and velocity approximations $U(Y, a_n(X))$ into the integral relations and subsequent integration leads to N ordinary differential equations of the form

$$\sum_{n=1}^{N} C_{n,k} \frac{da_n}{dX} = D_k \quad , \quad k = 1, 2, \cdots, N \tag{1}$$

where

$$C_{n,k} = \sum_{l=0}^{N} a_l\, P(k,l,n)$$

$$D_k = \frac{\dot{U}_e U_e}{\sigma_k} - \alpha \sum_{l=0}^{N} a_l - \dot{U}_e \sum_{l=0}^{N} a_l\, Q(k,l) - (V_0\, \sigma_k - \sigma_k^2) \sum_{l=0}^{N} a_l\, R(k,l)$$

Here $a_0(X) = U_e(X)$ and d/dX is replaced by a dot. P, Q and R are coefficients which are determined only once at the beginning of the calculation.

$$P(k,l,n) = 2[\sigma_k + (n+l)\alpha]^{-1} - 4[\sigma_k + (n+l+1)\alpha]^{-1} + 2[\sigma_k + (n+l+2)\alpha]^{-1}$$

$$+ \frac{\sigma_k}{n\alpha}\Big[[\sigma_k + (n+l)\alpha]^{-1} - [\sigma_k + l\alpha]^{-1} - [\sigma_k + (n+l+1)\alpha]^{-1} + [\sigma_k + (l+1)\alpha]^{-1}\Big]$$

$$+ \frac{\sigma_k}{(n+1)\alpha}\Big[[\sigma_k + l\alpha]^{-1} - [\sigma_k + (n+l+1)\alpha]^{-1} + [\sigma_k + (n+l+2)\alpha]^{-1} - [\sigma_k + (l+1)\alpha]^{-1}\Big]$$

$$Q(k,l) = 2[\sigma_k + l\alpha]^{-1} - 4[\sigma_k + (l+1)\alpha]^{-1} + 2[\sigma_k + (l+2)\alpha]^{-1}$$

$$+ \frac{\sigma_k}{\sigma_k + l\alpha}\Big[\alpha^{-1} - [\sigma_k + l\alpha]^{-1}\Big] + \frac{\sigma_k}{\sigma_k + (l+1)\alpha}\Big[[\sigma_k + (l+1)\alpha]^{-1} - 2\alpha^{-1}\Big]$$

$$+ \frac{\sigma_k}{\alpha}[\sigma_k + (l+2)\alpha]^{-1}$$

$$R(k,l) = [\sigma_k + l\alpha]^{-1} - [\sigma_k + (l+1)\alpha]^{-1}$$

The system (1) of first order ordinary differential equations is first solved for the rates of change da_n/dX which are then numerically integrated over X. The resulting a_n are used to calculate the velocity profile, wall shear, displacement thickness, and other quantities.

Nondimensional velocity:

$$U(X,Y) = (1 - e^{-\alpha Y}) \cdot [U_e(X) + \sum_{n=1}^{N} a_n(X)\, e^{-n\alpha Y} \tag{2}$$

Velocity: $u = u_\infty U$ [m/s]

Nondimensional wall shear stress:

$$T_w = \left.\frac{\partial U}{\partial Y}\right|_{Y=0} = \alpha\Big(U_e + \sum_{n=1}^{N} a_n\Big) \tag{3}$$

Wall shear stress: $\tau_w = (\rho \cdot u_\infty^2\, Re^{-1/2})\, T_w$ $\;[(\text{kg/m}^3)\,(\text{m/s})^2 = (\text{kg (m/s}^2)/\text{m}^2) = \text{N/m}^2]$

Nondimensional displacement thickness:

$$\Delta_1 = \int_0^\infty (1 - \frac{U}{U_e})\, dY = \frac{1}{\alpha U_e}\left[U_e + \sum_{n=1}^N a_n \left(\frac{1}{n+1} - \frac{1}{n} \right) \right] \tag{4}$$

Displacement thickness: $\delta_1 = (l\, Re^{-1/2})\, \Delta_1$ [m]

Simulation model

The simulation program was developed for $N = 3$ (three approximation terms) because for three simultaneous differential equations the solution of the system of equations (1) for the rates of change $ADOT$ ($= da_n/dX$) of the approximation parameters A ($= a_n$) can still be achieved without using subscripted variables (which requires sophisticated simulation software). Where such software is available, the program can be reformulated using matrices, and can then be used for higher order approximations producing even more accurate results. The solutions achieved with $N = 3$ are already highly accurate and differ only little from exact solutions. The independent variable *time* used internally in the simulation software is redefined as space coordinate X.

The simulation program for the computation of boundary layer flow over a circular cylinder with suction is listed completely in the following. The corresponding simulation diagrams are presented in Figures Z214a through c.

The flow velocity outside the boundary layer (UE) and the suction velocity (VB) are specified first. The constant coefficients $P(k, l, n)$, $Q(k, l)$ and $R(k, l)$ are calculated beforehand for $N = 3$ using standard programming software (like Visual Basic), and are then entered into the calculations for the coefficients of the system matrix $C_{n,k}$ and the sums appearing on the right hand sides D_k. Using Cramer's rule the rates of change $ADOT_n$ of the three state variables A_n are determined from the system matrix and the column vector of the right hand side. They are numerically integrated (4th order Runge-Kutta procedure RK4, time step $= 10^{-5}$). The A_n are then used to compute (with Equations 2 through 4) the local wall shear stress (TW), the displacement thickness of the boundary layer ($DELTA$ 1), and the local velocity profile (U).

Since the simulation software for ordinary differential equations provides for integration and presentation of results with respect to only one independent variable (usually *time*), the local velocity profiles cannot be continuously computed and stored. However, the computed time series for the approximation parameters a_n can be transferred to a spreadsheet program for computing the velocity profiles for an arbitrary location.

A simpler approach (same procedure as in model Z213) is taken here: At a location *Xprofil* (previously specified by the user) the rates of change $ADOT$ are set to zero. As the program continues running, the time steps (of *time*, redefined as X) are now redefined as steps in the Y-direction to allow computation of the velocity profile in this direction (U or $Urel = U/Ue$). For $Urel(Y)$ the velocity in the boundary layer is normalized with respect to the velocity (Ue) at the upper edge of the boundary layer (where $Urel = 1$). The velocity profile can be plotted by the simulation software as $U(Y)$.

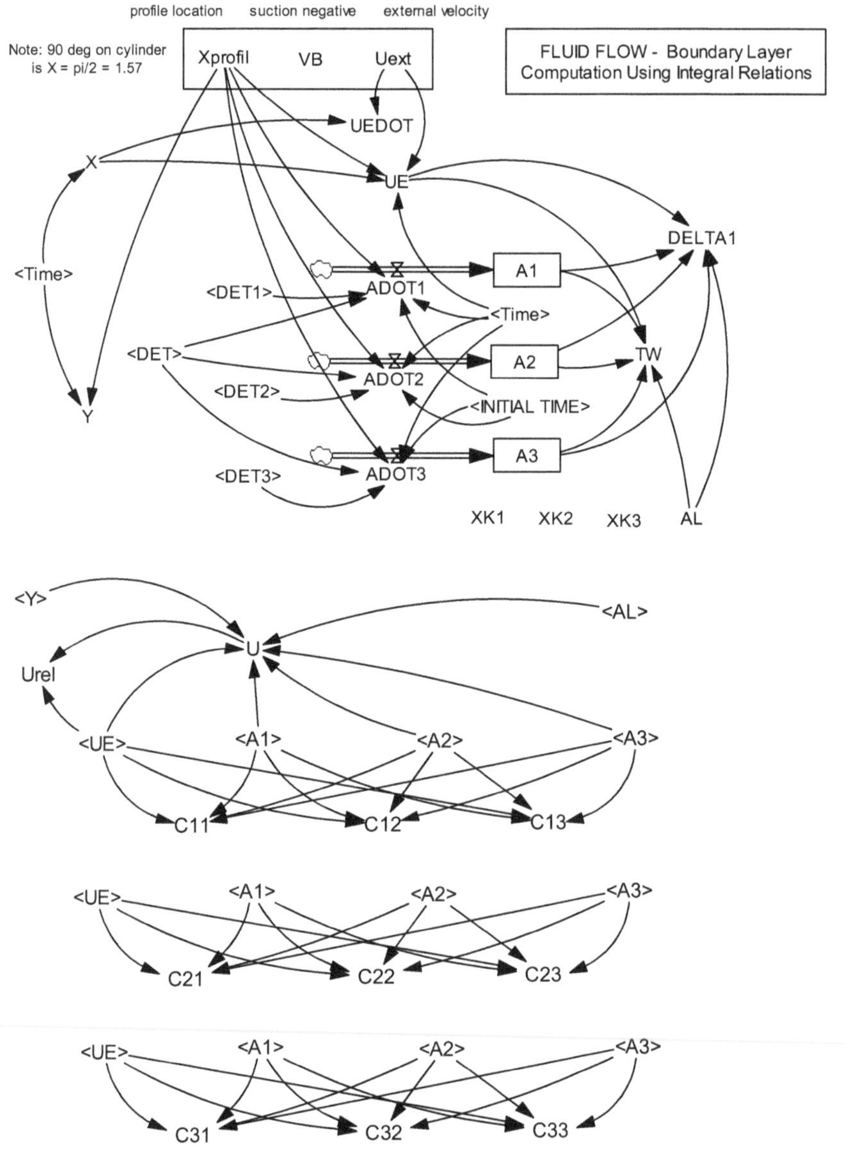

Figure Z214a: Simulation diagram for the boundary layer calculation – Part 1.

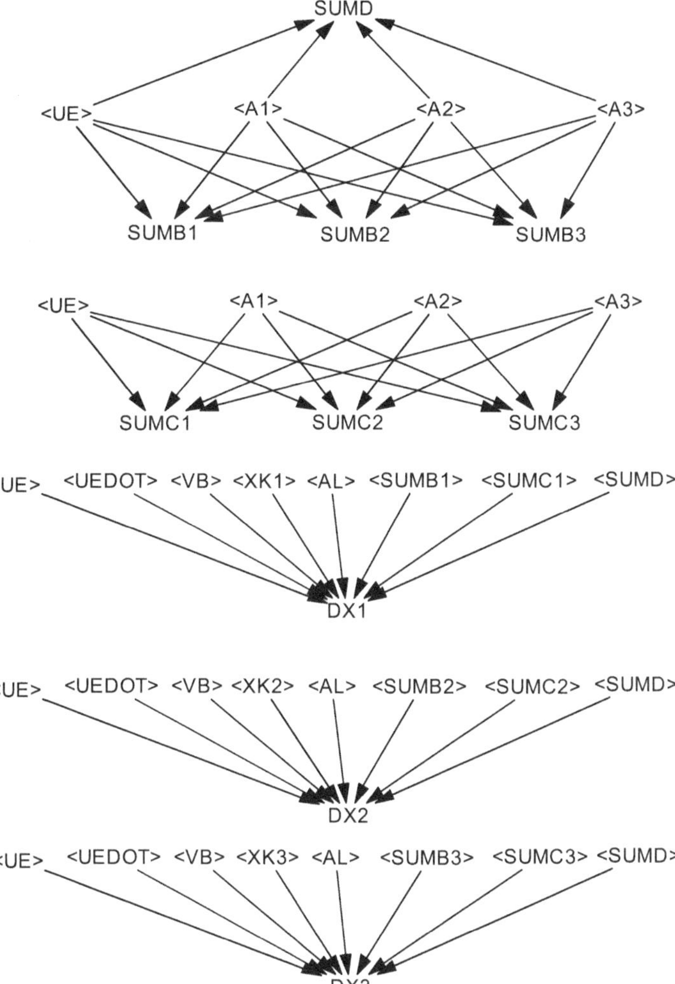

Figure Z214b: Simulation diagram for the boundary layer calculation – Part 2.

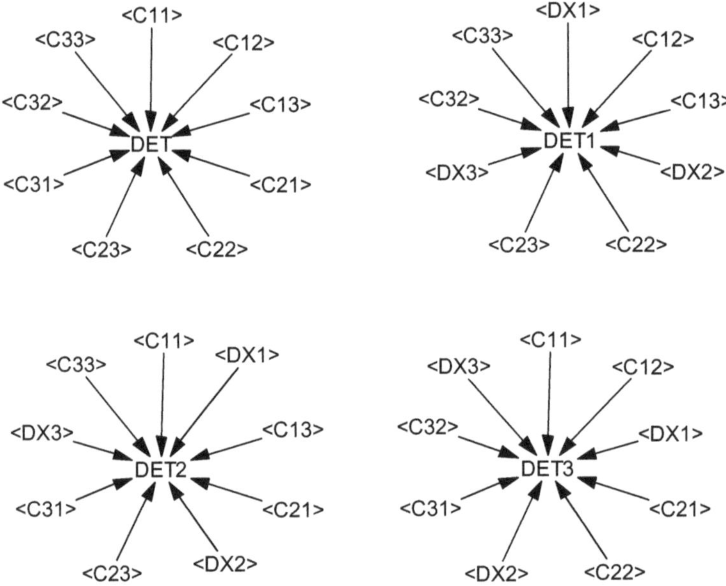

Figure **Z214c**: Simulation diagram for the boundary layer calculation – Part 3.

Parameters of the boundary layer simulation
(circular cylinder with external velocity distribution Ue = Uext sin (x/R), suction Vb)
Uext = 1 [1]
UE = IF THEN ELSE (Time < Xprofil, Uext *sin(X), Uext *sin(Xprofil)) [1]
UEDOT = Uext *cos(X) [1]
VB = -0.5 [1] *{suction is negative}*
Xprofil = 1.8 [1]
X = Time [1] *{the internal independent variable Time is used as x coordinate; x is in radian, i.e. x = pi/2 = 1.57 = 90 deg}*

Constants
AL = 1 [1]
XK1 = 1 [1]
XK2 = 2 [1]
XK3 = 3 [1]

Rates of change of velocity profile parameters
(when X = Xprofil is reached, the profile parameters are held constant to compute the velocity profile at this location)
ADOT1 = IF THEN ELSE ((Time<Xprofil) :AND: (Time>INITIAL TIME), DET1/DET, 0)
    [1]
ADOT2 = IF THEN ELSE ((Time<Xprofil) :AND: (Time>INITIAL TIME), DET2/DET, 0)
    [1]
ADOT3 = IF THEN ELSE ((Time<Xprofil) :AND: (Time>INITIAL TIME), DET3/DET, 0)
    [1]

Integration of profile parameters, with initial values (for stagnation point)
A1 = INTEG (ADOT1, 1.06658 *INITIAL TIME) [1]
     stagnation point profile: 1.06658, Blasius flat plate profile: -2.38035
A2 = INTEG (ADOT2, -1.32149 *INITIAL TIME) [1]
     stagnation point profile: -1.32149, Blasius flat plate profile: 3.60073
A3 = INTEG (ADOT3, 0.45108 *INITIAL TIME) [1]
     stagnation point profile: 0.45108, Blasius flat plate profile: -2.00504

Determinants for compution of the rates of change
DET = C11 *C22 *C33 +C12 *C23 *C31 +C13 *C21 *C32 -C13 *C22 *C31 -C12 *C21
     *C33 -C11 *C23 *C32 [1]
DET1 = DX1 *C22 *C33 +C12 *C23 *DX3 +C13 *DX2 *C32 -C13 *C22 *DX3 -C12
     *DX2 *C33 -DX1 *C23 *C32 [1]
DET2 = C11 *DX2 *C33 +DX1 *C23 *C31 +C13 *C21 *DX3 -C13 *DX2 *C31 -DX1
     *C21 *C33 -C11 *C23 *DX3 [1]
DET3 = C11 *C22 *DX3 +C12 *DX2 *C31 +DX1 *C21 *C32 -DX1 *C22 *C31 -C12
     *C21 *DX3 -C11 *DX2 *C32 [1]

Coefficients of the system matrix with factors P
C11 = UE/24 +A1/24 +A2/40 +A3/64.6154 [1]
C12 = UE/120 +A1/51.4286 +A2/72 +A3/105 [1]
C13 = UE*0 +A1/96.9231 +A2/118.588 +A3/160 [1]
C21 = UE/60 +A1/60 +A2/84 +A3/120 [1]
C22 = UE/180 +A1/114.546 +A2/140 +A3/184.39 [1]
C23 = UE/630 +A1/201.6 +A2/219.13 +A3/270 [1]
C31 = UE/120 +A1/120 +A2/152.727 +A3/201.6 [1]
C32 = UE/280 +A1/210 +A2/240 +A3/296.471 [1]
C33 = UE/672 +A1/347.586 +A2/360 +A3/420 [1]

Right hand side of differential equations
DX1 = -UEDOT*SUMB1 -(VB*XK1-XK1^2) *SUMC1 +UE*UEDOT/XK1 -AL*SUMD [1]
DX2 = -UEDOT*SUMB2 -(VB*XK2-XK2^2) *SUMC2 +UE*UEDOT/XK2 -AL*SUMD [1]
DX3 = -UEDOT*SUMB3 -(VB*XK3-XK3^2) *SUMC3 +UE*UEDOT/XK3 -AL*SUMD [1]

Partial sums with factors Q, R
SUMB1 = UE/4 +A1/9 +A2/19.4595 +A3/36.3636 [1]
SUMB2 = UE/18 +A1/27.6923 +A2/46.1538 +A3/73.2558 [1]
SUMB3 = UE/48 +A1/63.1579 +A2/91.3044 +A3/130.667 [1]
SUMC1 = UE/2 +A1/6 +A2/12 +A3/20 [1]
SUMC2 = UE/6 +A1/12 +A2/20 +A3/30 [1]
SUMC3 = UE/12 +A1/20 +A2/30 +A3/42 [1]
SUMD   = UE +A1 +A2 +A3 [1]

Wall shear stress and displacement thickness
TW = AL*(UE +A1 +A2 +A3) [1]
DELTA1 = (1/(AL*UE)) *(UE +A1*(1/2-1) +A2*(1/3-1/2) +A3*(1/4-1/3)) [1]

Velocity profile computation at X = Xprofil
Y = (Time-Xprofil)*10 [1]
U = (1 -EXP(-AL*Y))*(UE +A1*EXP(-AL*Y) +A2*EXP(-2*AL*Y) +A3*EXP(-3*AL*Y)) [1]
Urel = U/UE [1]

Simulation time parameters
INITIAL TIME = 0.01 [1]
FINAL TIME = 3 [1]
TIME STEP = 1e-004 [1 [0]
SAVEPER = 0.001 [1]

Simulation results

The results for laminar boundary layer flow over a circular cylinder with suction are shown in Figures Z214d to f. The flow over a circular cylinder has much in common with the flow around a wing profile – initial acceleration of the flow starting at the stagnation point (velocity zero), deceleration behind the point of maximum profile thickness, and finally separation of the flow and turbulence. The separation point – where the wall shear stress reduces to zero and forward flow appears in the boundary layer – also marks the end of the range for which the (parabolic) boundary layer equations produce a valid flow description. This is obvious also in the simulation results: The gradients become quite large in the immediate neighborhood of this point and further numerical calculation produces unreal results. In the case of suction, the flow separates near $x = 2$ (which is evident from the shear stress T_w approaching zero).

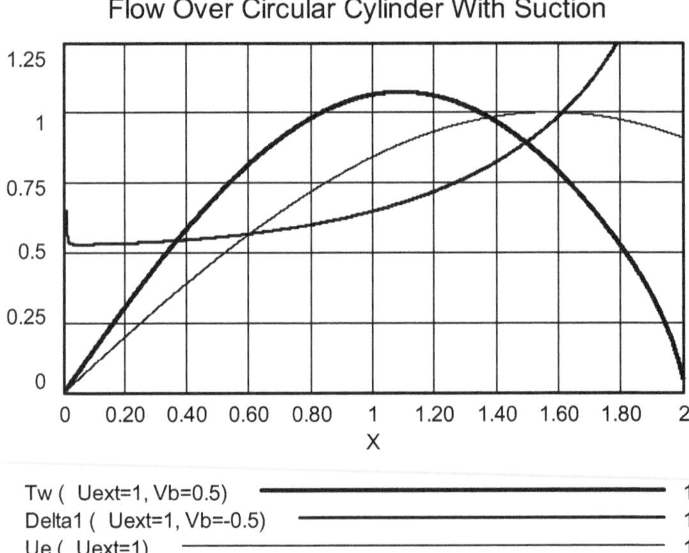

Figure Z214d: Wall shear stress, displacement thickness, and external velocity over the circular cylinder with suction.

The results (computed as nondimensional variables) must be converted to real variables using the relationships given above (Equations 2 to 4). The same nondimensional computation reveals, for example, that wall shear stress and boundary layer thickness will be much larger in a fluid of high viscosity (e.g. oil) than in a fluid of low viscosity (e.g. water).

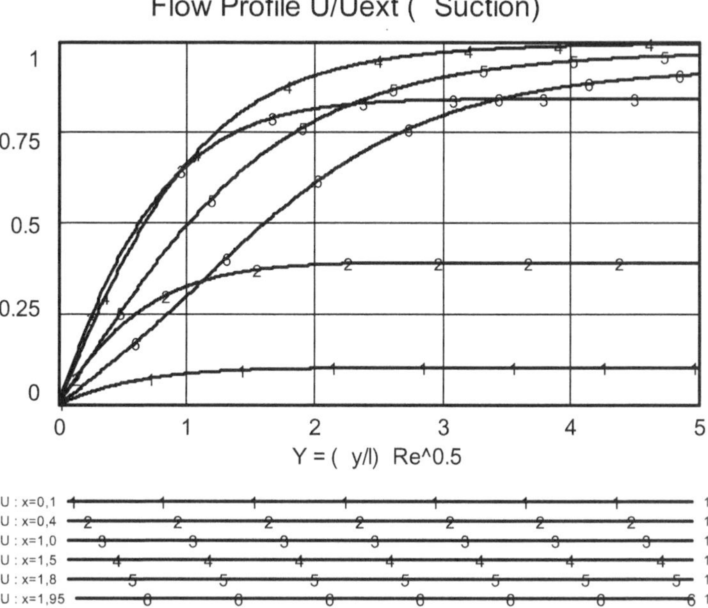

Figure Z214e: Velocity profiles in the boundary layer.

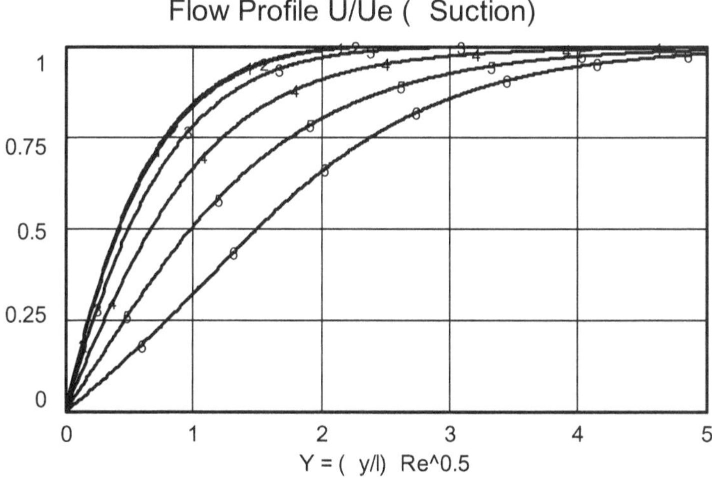

Figure Z214f: Velocity normalized to the velocity of the flow at the outer edge of the boundary layer.

Figure Z214d shows the (nondimensional) variables U_e, wall shear stress T_w, and displacement thickness Δ_1 as function of angle X (distance x on the surface of the cylinder divided by cylinder radius, $X = x/R$). The flow velocity outside the boundary layer is specified as $U_e = \sin X = \sin x/R$. It is zero at the stagnation point ($X = 0$) and has its maximum of $U_e = 1$ at the point of maximum thickness of the cylinder, i.e. for $X = \pi/2 = 1.57$ (for $R = 1$). This case has been calculated by several authors (Terrill 1960, Schönauer 1964, Bethel 1968) so that comparison becomes possible. The wall shear stress T_w has its maximum at $X \approx 1$, i.e. at about 60 degrees. After that it decreases, reaching zero at the separation point at $X \approx 2.0$ (approximately 115 degrees). The displacement thickness of the boundary layer increases continuously, in particular in the region behind the maximum profile thickness where the velocity U_e slows down again. The results agree extremely well with those of the other authors. If more approximation terms are taken into account, then the precision can be increased further (Bossel 1970 b).

Figure Z214e for the (relative) velocity (U/U_{ext}) shows how the flow accelerates quickly at the front of the cylinder from zero at the stagnation point to maximum velocity (at 90 degrees, i.e. $X = 1.57$), slowing down again behind the point of maximum thickness. In Figure Z214f the same velocity profiles are shown after normalization to the velocity U_e at the outer edge of the boundary layer (U/U_e). These normalized profiles show clearly that flow velocity near the wall slows down increasingly as the separation point is approached, causing the boundary layer to become thicker.

Figure Z214g shows results of a computation for the circular cylinder *without* suction ($V_b = 0$). In this case the boundary layer grows more quickly and separates earlier.

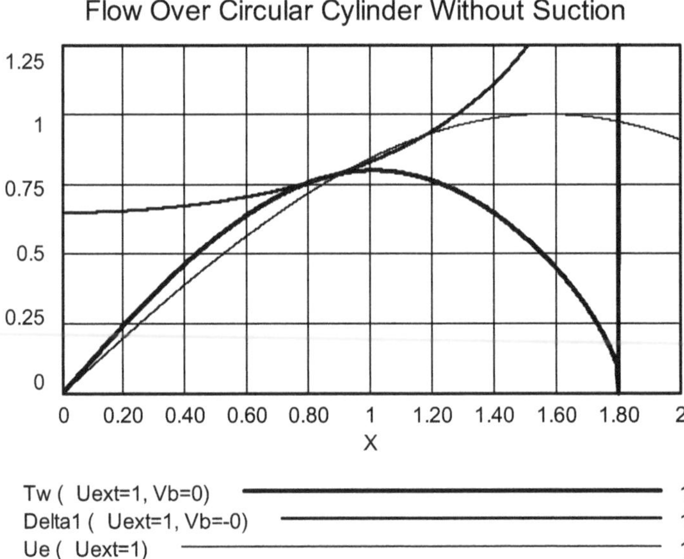

Figure Z214g: Wall shear stress, displacement thickness, and external velocity over the circular cylinder *without* suction.

Exercises

1. Examine the influence of computation step size (TIME STEP) on the "smoothness" of the results (using Runge-Kutta-4 integration). Compare the values for some key variables (like T_w, a_k) for the same X-locations for increasingly smaller step size. Can you identify convergence? Can you explain why the (numerical) error increases again as the step size is further reduced? Which step size do you recommend?

2. Vary the external velocity U_{ext} as well as the suction (or blowing) velocity V_b. What consequences does this have particularly for wall shear stress, displacement thickness, and point of separation?

3. Convert the nondimensional results to real flows, e.g. around a cylindrical column with $l = R = 2$ m in water or air flow of velocity $U_{ext} = 10$ m/s, applying the kinematic viscosity of (a) water (20°C): $v = 1 \cdot 10^{-6}$ [m²/s], (b) air (at 20°C): $v = 15 \cdot 10^{-6}$. Compute and compare the boundary layer profiles for $X = 1$.

4. Modify the program by prescribing corresponding external and suction velocities to compute other flows e.g. over a flat plate with discontinuous suction (cf. Bossel 1970b), or for a given velocity distribution over an airfoil. Try to find a velocity distribution (with and/or without suction) that moves the separation point as far to the rear edge as possible (increasing the ratio of laminar to turbulent flow, thereby reducing airfoil drag).

5. If simulation software allowing the use of subscripted variables is available, develop the present method for this software and compute several cases for $N = 2$ to 5 or 10. Check for convergence of results as N increases, and decide which N can be recommended for reliable results.

6. Implement the more complex procedure recommended in Bossel 1970b which uses an additional stretching function $g(X)$. Compute cases like those discussed above.

7. Investigate the influence of exponents α (*AL*) and σ_k (*XK1, XK2, XK3*) on results and convergence.

8. Clarify which additions and modifications are required to apply the procedure to vortex flows, jets, wakes, and compressible flows (cf. Bossel 1970a, 1970c, 1971, Mitra and Bossel 1971).

References

Bossel, H. 1969a: Vortex Breakdown Flowfield. *Physics of Fluids*, Vol. 12, No. 3, March 1969 (498-508).

Bossel, H. 1996b: Computation of Axisymmetric Contractions. *AIAA Journal*, Vol. 7, No. 10, October 1969 (2017-2020).

Bossel, H. 1970a: Use of Exponentials in the Integral Solution of the Parabolic Equations of Boundary Layer, Wake, Jet, and Vortex Flows. *Journal of Computational Physics*, Vol. 5, No. 3, 1970 (359-382).

Bossel, H. 1970b: Boundary Layer Computation by an N Parameter Integral Method Using Exponentials. *AIAA Journal*, Vol. 8, No. 10, October 1970 (1840-1845).
Note: The formula for P shown there on p. 1842 (bottom, right) should read:
$$P(k, l, n) = 2 [\sigma_k + \ldots$$

Bossel, H. 1970c: Vortex Computation by the Method of Weighted Residuals Using Exponentials. *AIAA Journal*, Vol. 9, No. 10, 1971 (2327-2334).

Mitra, N. K., Bossel, H. 1971: Compressible Boundary Layer Computation by the

Method of Weighted Residuals Using Exponentials. *AIAA Journal*, Vol. 9, No. 12, 1971 (2370-2377).

Bossel, H. 1971: Study of Vortex Flows at High Swirl by an Integral Method using Exponentials. *Lecture Notes in Physics*, Vol. 8, Springer-Verlag, New York 1971 (365-370).

Terrill, R. M. 1960: Laminar Boundary-Layer Flow Near Separation With and Without Suction. *Philosophical Transactions of the Royal Society*, London, Vol. A 253, Sept. 1960 (55-100).

Schlichting, H. 1960: *Boundary Layer Theory*. McGraw Hill, New York, 4th ed.

Bethel, H. E. 1968: Approximate Solution of the Laminar Boundary-Layer Equations with Mass Transfer. *AIAA Journal*, Vol. 6, No. 2, Feb. 1968 (220-225).

Schönauer, W. 1964: Ein Differenzenverfahren zur Lösung der Grenzschichtgleichung für stationäre, laminare, inkompressible Strömung. *Ingenieur-Archiv*, Vol. 33, 1964 (173-189).

Note: The description of model Z214 in the German edition *Systemzoo 1* (2004) contains an error in the formula for coefficient C21, producing incorrect results. The present formulation is correct.